DIGITAL CIRCUITS

DIGITAL CIRCUITS

J.R. NOWICKI

Senior Lecturer, Swindon College, Wiltshire

AND L.J. ADAM

Assistant Principal,
Blackburn College, Lancashire

Edward Arnold
A division of Hodder & Stoughton
LONDON MELBOURNE AUCKLAND

© 1990 J. R. Nowicki and L. J. Adam

First published in Great Britain 1990

Distributed in the USA by Routledge, Chapman and Hall, Inc.
29 West 35th Street, New York, NY 10001

British Library Cataloguing in Publication Data

Nowicki, J. R.
  Digital circuits.
  1. Digital intergrated circuits
  I. Title   II. Adam, L. J.
  621.381'73

  ISBN 0–7131–3641–3

Typeset in 10/11 pt Times by Colset Private Limited, Singapore
Printed and bound in Great Britain for Edward Arnold,
a division of Hodder and Stoughton Limited, Mill Road, Dunton
Green, Sevenoaks, Kent TN13 2YA by J.W. Arrowsmith Ltd,
Bristol

PREFACE

The rapid development of computers, microprocessors and related digital equipment has meant that an increasing number of people have become interested in understanding how these devices work. There is no longer the computer elite that existed a few years ago. The low price of personal computers has meant that the mysteries of computing have been revealed to a large number of non-specialist users. Microcomputers are in use in commerce, industry and the home. The use of computers is not confined to any one category of people – engineers of all persuasions use them for circuit design, numerical control of machine tools, process control, and in many other areas. Business people use computers for data storage, word processing, financial management, and so on. Teachers use them in schools to help children develop an understanding of basic concepts in an interesting way. Home computer 'buffs' and 'hackers' are now commonplace. Some become quite fanatical – perhaps initially attracted by the hypnotic computer games but eventually taking a deeper interest in the inner working of their machines. This usually develops into a desire to expand the system in some way, forcing an interest to develop in the hardware of the computer.

A major area of expansion in recent years has been the telecommunications industry, which is now largely dependent upon digital systems. The modern technologies of satellite systems and robotics are under intense development. The end result of this is that there is a continuing demand for expertise in the fields of digital systems and computing. This book is intended to provide a starting point for those interested in finding out about digital circuitry.

Understanding digital circuits involves the acquisition of a number of skills and an appreciation of the basic digital components and systems. An understanding of the binary number system is essential, and this is fully explained in Chapter 2 together with details of the octal and hexadecimal number systems that are closely related to binary. In addition, Chapter 2 introduces a special code used in digital systems for handling decimal number representation.

Logical functions, relating logic circuit inputs and outputs, are described and manipulated using Boolean Algebra, a special form of algebra that allows symbolic representation of logic levels. This is explained in detail in Chapter 3. Simple logic gates are introduced in Chapter 4, which also contains examples of combinations of these gates to form circuits that are of general use. Specific circumstances dictate particular needs, and a study of this chapter will provide the reader with the ability to design any non-time-dependent logic circuit to fulfil a specialist requirement. This could be anything from a simple interlock circuit to a memory decoding circuit for a microprocessor system. When circuit requirements become more complex, it is necessary to try to keep the number of components to a minimum. This is not because the cost of components is a major factor – if less components are used, power consumption and power dissipation are reduced, the number of connections to be made is less and the manufacturing process is simplified. In addition, the simpler a circuit is, the less likely (in general) it is to go wrong. Chapter 5 explains accordingly the various minimisation techniques available. Chapter 6 investigates the electrical aspects of digital circuitry and describes the major logic families available to the circuit designer, with the associated electrical implications. The assumption is made in Chapter 6 that the reader has some familiarity with simple transistor and associated electronics.

Temporary storage is of vital importance in digital systems, and Chapter 7 introduces the flip-flop or bistable, a device which can store a 0 or a 1. Registers, which are devices to store groups of 0s and 1s, are described in detail in Chapter 8, and Chapter 9 introduces the many types of counters that are used to produce particular sequences of patterns of 0s and 1s. Chapter 10 deals with logic circuits that perform arithmetic functions including addition, subtraction, multiplication and division.

There are many options available to the circuit designer when implementing a design. Chapter 11 deals with methods of implementation of digital circuits and includes details of logic arrays and multiplexers. Chapter 11 also introduces digital fault finding techniques and equipment. Chapter 12 examines the way in which bit patterns are used to represent a variety of quantities or operations including numbers, memory addresses and operations in computers and analogue equipment, illustrating the range of codes available for the various applications. Many coding circuits are introduced as examples of applications of the techniques of combinational and sequential logic design introduced in the text.

Appendix A gives some design exercises which can all be implemented using integrated circuits and which are suitable for laboratory exercises. Appendix B details advanced minimisation techniques, and Appendix C describes a practical circuit for a 12 or 24 hour clock which can be implemented using simple integrated circuits.

Swindon, 1989

CONTENTS

1

DIGITAL CIRCUITS AND THEIR APPLICATIONS

1.1 Introduction

In **analogue** electronic circuits, (sometimes called **linear circuits**), voltage levels can vary continuously. An example of this is a transistor amplifier which can amplify any voltage level within a specified range. **Digital** circuit voltage levels, however, are restricted to values which are predetermined. In most digital circuits there are only two levels, for example 0 V and 5 V. These voltage levels are referred to as **logic levels**. With 2–state logic these levels are referred to as either 0 or 1. By grouping 0s and 1s together, digital information can be monitored, modified or stored. This is the basis of operation of the many thousands of types of digital integrated circuits that are now readily available at low cost. This includes devices known as gates, bistables or flip-flops, registers, counters, multiplexers and microprocessors.

1.2 Electrical Aspects

It has already been explained that digital circuits generally operate on only two voltage levels. This will be assumed throughout this text. These voltage levels are not standard, however, and different forms of digital circuit (which are manufactured using different processes) may not be directly compatible. This is not an intentional ploy by circuit manufacturers to make life difficult – there are very good reasons for the differences. This

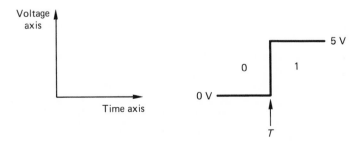

Fig 1.1 Positive-going transition with zero rise time

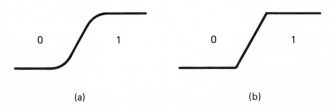

(a)   (b)

Fig 1.2   Positive-going transition with appreciable rise time

will become clear when Chapter 6 has been tackled. For the moment, however, particular voltage levels will be assumed. A system representing logic level 0 as 0 V and logic level 1 as 5 V is illustrated in Fig. 1.1.

This shows a waveform representation of logic levels at a **node**, or point in a circuit, that changes from 0 V (logic 0) to 5 V (logic 1) at time T. This is referred to as a **Positive-going transition**; alternatively, the transition at time T can be referred to as a **leading edge**. In reality it is not possible to change a voltage from one value to another instantaneously (zero rise time), and a more realistic impression is given in Fig. 1.2(a). For this reason transitions are often represented as in Fig. 1.2(b). Note that there is a delay, which means that the voltage takes time to rise to its final level.

If transitions are excluded, only two voltage levels are possible, so a node in a circuit must be at one or other of these levels at any particular time. This **static** logic level could be checked with a d.c. voltmeter or with an oscilloscope. A sequence of voltage levels would be detected by the voltmeter as a succession of high and low voltages or by the oscilloscope, which traces out the variation of voltage with respect to time, as a waveform showing a series of positive and negative edges. At useful frequencies the d.c. voltmeter would be incapable of following the rapidly changing voltage levels, and an oscilloscope would then be essential. A typical waveform is represented in Fig. 1.3.

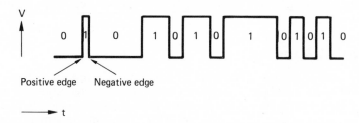

Fig 1.3   Waveform display on osculloscope

The sequence of 0s and 1s represented in Fig. 1.3 corresponds to **coded** information, for example it could represent a number or a letter of the alphabet. It is a common requirement to transmit such information from one part of a digital system to another – for example a digital computer may require to print the letter 'A' on a printer. One method is to connect the computer to the printer via a **serial** link. This is illustrated in Fig. 1.4.

If the voltage on the computer output (CO) is made to vary with respect to time, as previously shown in Fig. 1.3, the connecting wire will ensure that the same variation takes place at the printer input (PI). If the variation follows a sequence which is the code for the

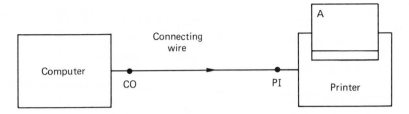

Fig 1.4 Serial Link between computer and printer

letter 'A', the digital circuitry within the printer will recognise it and cause the 'A' to be printed.

The information, or **data**, that has been transmitted (i.e, the letter 'A') has been sent in a **serial mode**. The great advantage of serial transmission is that only one transmission line is required. The disadvantage is that, as only one voltage level can be sent at any one instant, serial transmission is relatively slow. An alternative method is to use **parallel transmission** as shown in Fig. 1.5.

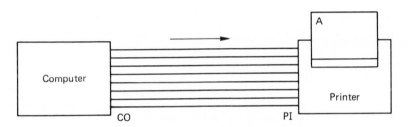

Fig 1.5 Parallel data transmission

With this method, if it is assumed that eight voltage levels are used to define the letter 'A' uniquely then the printer input will receive all the appropriate levels as soon as they are presented at the computer output. Clearly this is a faster method of data transmission than the serial case. The disadvantage of parallel transmission is that a greater number of lines is required, with the resulting penalties of cost and increased construction time.

Generally speaking, parallel transmission is used over short distances and serial transmission over long distances where the extra lines would be too expensive. It will of course be required to send further characters to be printed in the example of Fig. 1.5, and all that is necessary is to present the codes in sequence at the computer output, to be received and printed in the same sequence at the printer. An oscilloscope could then be used on each line to show the sequence of voltage levels. The waveforms could be as given in Fig. 1.6 which shows all eight waveforms simultaneously. This shows the changing voltage levels on each line for the transmission of the characters A, B, C, D, E, F, G, H, I to the printer.

In both the serial and parallel transmission examples illustrated, some form of accompanying timing signals would be required to regulate the points at which a new voltage level is transmitted and received.

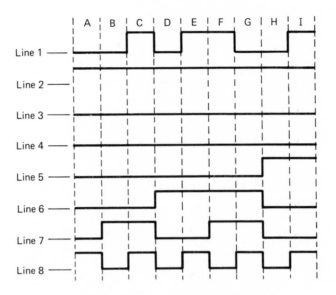

Fig 1.6 Sequence of parallel characters

1.3 Applications of Digital Circuits

Control Applications

A major application of digital circuits is in the area of process control. A simple example would be a safety device for a motor vehicle. It may be a requirement that the engine will only be allowed to start under certain conditions. These conditions could be:

(a) The driver's door is closed.
(b) The driver must be wearing a seat belt.
(c) The bonnet must be fastened.

In this simple example each requirement can be defined in two-state terms, i.e. the driver's door will be open or closed, the seat belt will be fastened or unfastened, and the bonnet must be fastened or unfastened. This is illustrated in Fig. 1.7.

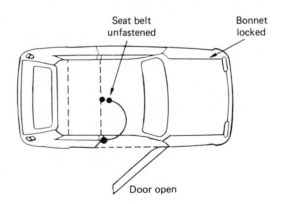

Fig 1.7 Simple control example

Table 1.1   Tabulation of all possible inputs

Input devices			Output signal
Door	Bonnet	Seat belt	To engine
0	0	0	0
0	0	1	0
0	1	0	0
0	1	1	0
1	0	0	0
1	0	1	0
1	1	0	0
1	1	1	1

A microswitch could be used to detect whether the door is open or closed. If the door is closed the microswitch would generate, say, 5 V to feed to the control circuit. If the door is open, the switch would provide 0 V. If 5 V is assumed to be logic 1 and 0 V is assumed to be logic 0, then a 1 indicates a closed door and a 0 an open door. Similarly a contact in the seat belt fastening could produce a 1 to indicate a fastened seat belt and a 0 to indicate an unfastened seat belt. A fastened bonnet would be a 1 and an unfastened bonnet a logic 0. Only when all of these conditions are satisfied would the engine be allowed to start. The control circuit will only generate a 5 V output, representing permission to start the engine, in that case. A 1 therefore signifies 'enable engine start' and a 0 output 'inhibit engine start'. This is summarised in Table 1.1 above.

Note that only when the door is closed AND the bonnet is locked AND the seat belt is fastened ($D = B = S = 1$) will the engine be allowed to start ($E = 1$). A block diagram representing the circuit requirement is shown in Fig. 1.8.

This is an easy task for a digital circuit. The important point to note at this stage is the simple way in which the problem can be defined using logic levels 0 and 1. More complex control problems could be defined in the same way.

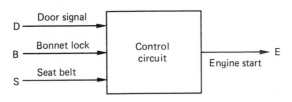

Fig 1.8   Circuit requirement

Example 1.1

A conveyor belt for a production line at a bottling plant must only be allowed to start under the following conditions:

(a) Sufficient empty bottles are available at the start of the line.
(b) The liquid in the bottle filler is higher than a minimum level.
(c) Sufficient bottle caps are stored in the capping machine hopper.
(d) Sufficient labels are available in the labelling machine.
(e) Alternatively an override switch has been set.

Produce a table representing all possible inputs and the corresponding outputs.

Solution

>B = 1 means sufficient bottles.
>F = 1 means liquid > minimum level.
>C = 1 means sufficient caps.
>L = 1 means sufficient labels.
>S = 1 means override switch set.

The combinations of input values and resulting outputs are shown in Table 1.2.

Table 1.2   Solution to Example 1.1

B	F	Inputs C	L	S	Output	B	F	Inputs C	L	S	Output
0	0	0	0	0	0	1	0	0	0	0	0
0	0	0	0	1	1	1	0	0	0	1	1
0	0	0	1	0	0	1	0	0	1	0	0
0	0	0	1	1	1	1	0	0	1	1	1
0	0	1	0	0	0	1	0	1	0	0	0
0	0	1	0	1	1	1	0	1	0	1	1
0	0	1	1	0	0	1	0	1	1	0	0
0	0	1	1	1	1	1	0	1	1	1	1
0	1	0	0	0	0	1	1	0	0	0	0
0	1	0	0	1	1	1	1	0	0	1	1
0	1	0	1	0	0	1	1	0	1	0	0
0	1	0	1	1	1	1	1	0	1	1	1
0	1	1	0	0	0	1	1	1	0	0	0
0	1	1	0	1	1	1	1	1	0	1	1
0	1	1	1	0	0	1	1	1	1	0	1
0	1	1	1	1	1	1	1	1	1	1	1

Study Table 1.2 to confirm an understanding of the solution. Using this technique, digital circuits can be designed to control any industrial process. Consider the case of a nuclear power station with its comprehensive interlock systems to ensure safety. Digital circuits can monitor a variety of parameters such as reactor temperature and pressure, control rod positions, and so on. Based on these parameter values, the circuit can generate appropriate control signals. Another example would be a railway signalling system which detects train positions and sets signals accordingly.

Digital Computers and Microprocessor systems

Digital computers have now become a part of everyday life. They are used in offices, banks and institutions of all kinds, in aircraft and in the home. They have endless applications, limited only by the imagination of the human mind. Their great versatility arises from the fact that they are **programmable**. Programs of instructions, or **software**, can be loaded to suit the particular application. These include control applications such as those considered above, so that the problem of Example 1.1 could be solved by a digital computer using a special program.

Example 1.2
Demonstrate how a program could be devised to solve the problem of Example 1.1 using a digital computer.

Solution

This can best be illustrated by means of a **flowchart**, which is a diagrammatic representation of an algorithm solving the problem.

This is given in Fig. 1.9.

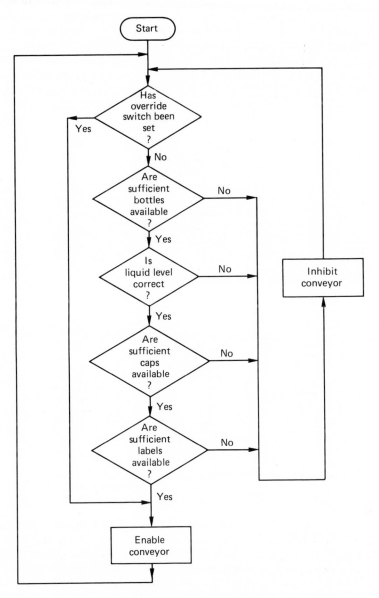

Fig 1.9 Flowchart for example 1.2

The next step would be to convert the flowchart into program instructions in the same sequence.

There are therefore two distinct methods of solving the same problem. The first (hardware) method is to design and build a circuit specifically to solve this problem. The circuit cannot be used for anything else. The second (software) method would be to produce a suitable program to run on a digital computer connected via appropriate interfacing circuits to the system. The advantage of the second method is that should system requirements change, it would be a simple matter to rewrite the program to suit. However, the system response time may be slower with the computer than with the first method, and the computer will probably cost more; but the flexibility of the computer may outweigh these disadvantages. It must be realised that the computer itself is simply a collection of digital circuits. These are an excellent example of digital technology, and they will be used to illustrate typical circuits. To begin with, a block diagram of a microcomputer is given in Fig. 1.10.

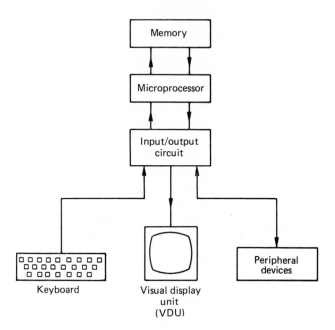

Fig 1.10   Microcomputer block diagram

The heart of the microcomputer is the microprocessor. The processor of any computer is the part in which arithmetic and logical operations are carried out, and is the source of control signals for the rest of the microcomputer. The microprocessor is a processor normally contained in a single integrated circuit (IC). This is illustrated in Fig. 1.11(a). Integrated circuits are frequently mounted in a package of the type shown in Fig. 1.11(a). The 'chip' itself is mounted centrally and connected to the pins via fine connecting wires as shown in Fig. 1.11(b).

The chip contains all of the electronic components – transistors, diodes, resistors – that are needed to implement the circuit function. In the case of the microprocessor, vast numbers of components are put onto a single chip of silicon. A microprocessor is a good example of a digital device that operates in parallel. A typical microprocessor will send or

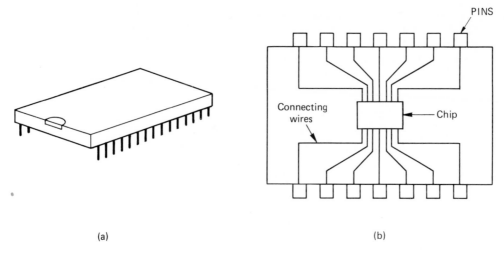

Fig 1.11 (a) Microprocessor integrated circuit; (b) integrated circuit construction

receive information in groups of eight or 16 lines called **buses**. The microprocessor hardware consists of registers, counters, gates, etc. – in fact all of the types of device described in this book. Although it is unlikely that the reader will ever design a microprocessor, a knowledge of the system components and their operation will greatly enhance the understanding of microprocessor operation.

In Fig. 1.10 the block marked 'Memory' will in almost all cases be in integrated circuit form. Memory chips store information in groups of typically 8 or 16 binary digits (0s and 1s or **bits** in abbreviated form). Data is interchanged between the memory and microprocessor in parallel. The input/output circuit will also consist of digital integrated circuits, with facilities for linking external devices such as the keyboard, visual display unit and other devices to the microprocessor, as shown in Fig. 1.10. These circuits will perform tasks such as parallel to serial conversion, temporary data storage, error detection and correction and other tasks specific to the peripheral device concerned. Many of the circuits, devices and techniques covered in this book are used in digital computer circuits and associated equipment.

1.4 Problems

1.1 Explain the difference between a linear circuit and digital circuit.

1.2 Explain how a logic level is physically represented.

1.3 How can logic levels be used to represent information?

1.4 Do all logic level representations adhere to common voltage standards?

1.5 Can voltage levels change instantaneously from one logic level to another?

1.6 How can a circuit node be checked to determine a static logic level?

1.7 The sequence 10111101 is transmitted serially down a connecting link between a digital computer and a printer. Sketch the waveform that would be observed on an oscilloscope connected to the serial link.

1.8 What is the advantage of parallel transmission?

1.9 An electric motor driving a pump will only be allowed to start if the temperature of the windings is less than 60° C, if there is liquid in the pump and if a start switch is operated. Represent this requirement in tabular form using logic level representation. Also produce a flowchart to show how the problem could be solved using software.

2

NUMBER SYSTEMS

2.1 Introduction

Chapter 1 explains that digital circuits and systems generally operate on only two levels of logic, represented by 0 and 1, and that combinations of 0s and 1s can be used to represent information in a coded form. There is an important difference to be noted here between a pattern of 0s and 1s representing a number and a pattern of 0s and 1s representing other information such as the character 'A' or the character '?'. If it represents a number, then in most cases each 0 or 1 has a value depending upon its position in the number, in the same way that the '7' in the decimal number '79' represents $7 \times 10 = 70$ because of its position. From a knowledge of this, the value of the number can be calculated. Numbers using 0s and 1s only are called **binary numbers**.

Numbers can be represented in a variety of number systems, depending upon the base adopted. The number system most generally used in everyday work is the **decimal** number system which has a base of ten. Unfortunately, the decimal number system does not lend itself to easy representation in digital circuits as ten voltage levels would be needed to represent the digits 0 to 9. (However, there is a frequent need to convert from binary numbers to decimal numbers and *vice versa* and it is important to be able to do this.) As the binary number system uses only two symbols to represent all values it is said to have a **base** or **radix** of $2^1 (= 2)$. Another number system which is related to binary is the **octal** number system, which uses eight symbols and has a radix of $2^3 (= 8)$. The **hexadecimal** number system, again related to binary, has sixteen symbols and a radix of $2^4 (= 16)$.

Because sixteen is two raised to the power of four, and two is two raised to the power of one, there is a direct relationship between the hexadecimal and binary number systems. A similar relationship exists between the octal and binary number systems. Direct conversion between these systems is very easy, and binary numbers can be represented in shorthand form by either octal or hexadecimal values. Large binary numbers are very easy to misread, and 0s and 1s can very easily be interchanged or lost. The techniques for conversions between these number systems are described in this chapter.

On many occasions patterns of 0s and 1s do not represent binary numbers. Examples of these are given in Chapter 1, where an alphabetic character (e.g. 'A') and combinations of input devices in a car safety system are represented by such patterns. There are many occasions when non-numeric patterns arise and there are a large number of codes used for a variety of applications. Binary codes are used to represent letters, full stops, commas, decimal numbers in calculators and petrol pump displays, angular positions in shaft encoders, and so on. As this chapter is concerned primarily with numbers, only one of these, the binary coded decimal code used to represent decimal digits, will be mentioned. Further work on codes will be covered in Chapter 12 at which stage coding circuits can also be considered. As the decimal number system is so familiar, it will be the first to be dealt with.

2.2 Decimal numbers

The decimal number system has *ten* symbols. These are:

0 1 2 3 4 5 6 7 8 9

Note that there are ten unique symbols, including zero, and that there is no symbol for ten. To obtain ten, two of the existing symbols in the set must be combined, ie. 1 and 0. It is a feature of most number systems that the base or radix is not uniquely represented by a symbol. To represent in decimal larger numbers than 9, a system of **weighting** must be used. For example, the number 896 really means:

$$(8 \times 100) + (9 \times 10) + (6 \times 1)$$

The position of the digit determines its weighting. The weighting is the multiplying factor that is applied to the digit. It can also be given in the following form:

$$896 = (8 \times 10^2) + (9 \times 10^1) + (6 \times 10^0)$$

Thus for the decimal system, the weighting follows the sequence:

$$\ldots 10^3 \quad 10^2 \quad 10^1 \quad 10^0 \quad . \quad 10^{-1} \quad 10^{-2} \quad 10^{-3}$$

whole numbers   fractions
or integers

Thus the decimal system will follow the number sequence:

	0	1	2	3	4	5	6	7	8	9 ← Run out of symbols so
First repeat →	10	11	12	13	14	15	16	17	18	19 re-use them with a prefixed repeat number
Second repeat ↗	20	21	22	23	24	25	26	27	28	29
	30	. . . etc.								

To signify a decimal number, a subscript is added as shown below:

$$896_{10}$$

2.3 The binary number system

An early attempt by Charles Babbage to produce a mechanical decimal computer in 1834 was a failure. Representing ten symbols in a machine – mechanical or electronic – is rather complex, and in fact is not necessary. It is much simpler to represent only two symbols and it is on this basis that digital systems, computers and microprocessors operate.

The symbols are:

0   1

There are only two unique symbols, including zero, and there is no unique symbol for two. To obtain '2', two of the existing symbols are combined to produce it.

In electrical terms, the symbols 0 and 1 can be represented by two voltage levels, e.g. $1 = 5\,V$, $0 = 0\,V$. Thus a waveform as shown in Fig. 2.1 represents a series of 0s and 1s.

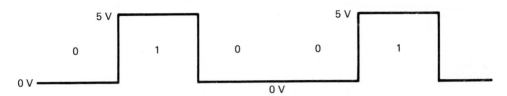

Fig 2.1   Typical waveform

Weighting also applies to the binary number system. Thus 11001 means:

$$(1 \times 2^4) + (1 \times 2^3) + (0 \times 2^2) + (0 \times 2^1) + (1 \times 2^0)$$
$$= 25_{10}$$

The sequence of weightings for binary numbers is as follows:

Binary (radix) point

$$\ldots 2^3 \quad 2^2 \quad 2^1 \quad 2^0 \quad . \quad 2^{-1} \quad 2^{-2} \quad 2^{-3}$$
$$8 \quad 4 \quad 2 \quad 1 \quad . \quad \tfrac{1}{2} \quad \tfrac{1}{4} \quad \tfrac{1}{8}$$

Decimal to binary conversion is a frequent exercise that has to be dealt with when using digital systems. It is a good idea to memorise the binary and decimal equivalents up to 15_{10}.

Binary	decimal	Binary	decimal
0000	0	1000	8
0001	1	1001	9
0010	2	1010	10
0011	3	1011	11
0100	4	1100	12
0101	5	1101	13
0110	6	1110	14
0111	7	1111	15

To convert larger numbers, divide the decimal number by 2 and keep the remainder. For example, to convert 25_{10} to binary:

2	25		Least significant bit (LSB)
2	12	r 1 ✓	(smallest weighting)
2	6	r 0	
2	3	r 0	
2	1	r 1	
	0	r 1	← Most significant bit (MSB) (largest weighting)

Thus $25_{10} = 11001_2$.

Note that a binary number is signified by a subscript 2.

2.4 Bits, bytes and words

It is a good opportunity now to define some terms associated with binary numbers and their use before proceeding to some worked examples of decimal to binary conversion.

The standard way of referring to an 0 or 1 is to use the term **bit**. The word 'bit' is derived by abbreviating binary digit.

A **byte** is a group of eight bits, for instance

$$0\ 1\ 1\ 0\ 1\ 0\ 0\ 1_2$$

An eight-bit grouping is extremely common, so it is essential to be familiar with bytes.

A less common grouping of four bits is referred to as a **nibble**.

Although there is some standardisation between computers and digital systems in how the hardware is organised to deal with particular groupings of bits, a general expression is used to refer to the number of bits decided upon for a particular machine. This is the **wordlength** of the system. It is normally fixed for a specific machine but can, in general, be of any length. Common lengths of words in computer systems are 8, 16 and 32. For instance, a computer word of length 16 bits could be:

$$0\ 1\ 1\ 1\ 1\ 0\ 0\ 0\ 1\ 0\ 0\ 1\ 0\ 1\ 1\ 1$$

Note that this 16-bit word corresponds to two bytes or four nibbles. Clearly the handling of such a large word is difficult for us to deal with as humans, and so the octal and hexidecimal systems are used to make this easier. These systems are fully described later in the chapter.

Another 16-bit word could be

$$0\ 0\ 0\ 1\ 0\ 0\ 0\ 0\ 0\ 0\ 0\ 1\ 1\ 0\ 0\ 1$$

↑   ↑

MSB   LSB

The terms MSB and LSB mean most significant bit and least significant bit respectively, and refer to the bits with greatest and least weighting respectively. The same 16-bit word could be rewritten:

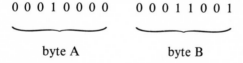

0 0 0 1 0 0 0 0   0 0 0 1 1 0 0 1

byte A   byte B

The word splits into two bytes. Byte A is the most significant byte, and byte B is the least significant byte.

The terms integer, fraction and mixed number also need definition. An **integer** is a whole number. Integers in any number system always have the radix point immediately to the right of the least significant digit. A **fraction** lies on the other side of the radix point, and is simply a number that is less than 1. A **mixed number** is a combination of an integer and a fraction.

Another useful term to have ready for use is register. A **register** is a hardware device for storing a word. Hardware details will be fully covered later in this text.

Example 2.1

Convert the following decimal numbers to their binary equivalent.
(a) 37   (b) 112   (c) 127

Solution

(a) 2 | 37

2 | 18 | r 1   LSB

2 | 9 | r 0

2 | 4 | r 1

2 | 2 | r 0

2 | 1 | r 0

0 | r 1   MSB

Thus

$37_{10} = 100101_2$

Check

100101

$32 + 4 + 1 = 37_{10}$

(b) 2 | 112

2 | 56 | r 0   LSB

2 | 28 | r 0

2 | 14 | r 0

2 | 7 | r 0

2 | 3 | r 1

2 | 1 | r 1

0 | r 1   MSB

Thus

$112_{10} = 1110000_2$

Check

1110000

$64 + 32 + 16 = 112_{10}$

(c) 2 | 127

2 | 63 | r 1   LSB

2 | 31 | r 1

2 | 15 | r 1

2 | 7 | r 1

2 | 3 | r 1

2 | 1 | r 1

0 | r 1   MSB

Thus

$127_{10} = 1111111_2$

Check

1111111

$64 + 32 + 16 + 8 + 4 + 2 + 1 = 127_{10}$

Note that in (c) the solution is seven 1s. This is because 127 is 1 less than 128, which equals 2^7. Any number that is 1 less than a power of 2 consists of all 1s. Powers of 2 consist of a 1 followed by **n** zeros where n is the power, e.g. $2^7 = 128_{10} = 10000000_2$

It is worth remembering this fact because of the frequency with which it is necessary to determine powers of 2 when using digital systems, particularly computers and microprocessors.

2.5 Powers of 2

If the intention is to become fully conversant with digital systems then it is a good idea to commit to memory powers of 2 up to and including 2^{16} as follows:

Powers of 2	Decimal equivalent	Powers of 2	Decimal equivalent
0	1	9	512
1	2	10	1024
2	4	11	2048
3	8	12	4096
4	16	13	8192
5	32	14	16384
6	64	15	32768
7	128	16	65536
8	256		

The reason for choosing 16 as a maximum is that 16 is a common wordlength in computers and microprocessors.

2.6 Binary fractions

The above examples of course all referred to integer conversion from decimal to binary. The process for integer decimal to binary conversion consists of repeated division by 2. Fractional decimal to binary conversion consists of repeated multiplication by two, as the following examples will illustrate.

Example 2.2
Convert the following decimal fractions to binary:
(a) 0.0625   (b) 0.11   (c) 0.6

Solution

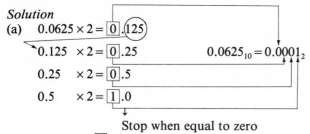

(a) $0.0625 \times 2 = \boxed{0}.(125)$

$0.125 \times 2 = \boxed{0}.25$ $0.0625_{10} = 0.0001_2$

$0.25 \times 2 = \boxed{0}.5$

$0.5 \times 2 = \boxed{1}.0$

Stop when equal to zero

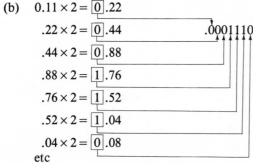

(b) $0.11 \times 2 = \boxed{0}.22$

$.22 \times 2 = \boxed{0}.44$ $.0001110$

$.44 \times 2 = \boxed{0}.88$

$.88 \times 2 = \boxed{1}.76$

$.76 \times 2 = \boxed{1}.52$

$.52 \times 2 = \boxed{1}.04$

$.04 \times 2 = \boxed{0}.08$

etc

The further the conversion process is taken, the more accurate the conversion will be. In practice, a limit will be set by the wordlength of the system in use. A longer wordlength will always give a more accurate representation unless conversion is complete before the limit imposed by the wordlength.

(c) $0.6 \times 2 = 1.2$
    $.2 \times 2 = 0.4$
    $.4 \times 2 = 0.8$
    $.8 \times 2 = 1.6$
    $.6 \times 2 = 1.2$
    $.2 \times 2 = 0.4$

This is beginning to look familiar. Clearly the pattern is going to repeat to infinity, and so an exact representation is not possible in binary. The above solution gave $0.6_{10} = 0.10011_2$ to 5 places. It is worthwhile converting this back to decimal to determine the error.

0.10011_2 is equivalent to $0.5 + 0.0625 + 0.03125$

0.5
0.0625
0.03125
_____

0.59375   which gives a considerable error.

Mixed decimal to binary conversion is simply a process of combining the integer and fractional techniques outlined above.

2.7 Binary arithmetic

Binary arithmetic will now be considered. To be able to deal with circuitry to carry out arithmetic processes such as addition, subtraction, multiplication and division, it is first necessary to examine the processes themselves.

Addition

First of all, the rule for adding binary numbers:

$0 + 0 = 0$ carry 0
$0 + 1 = 1$ carry 0
$1 + 0 = 1$ carry 0
$1 + 1 = 0$ carry 1

Note that you should not confuse the addition sign with the logical sum (OR) symbol, which is the same.
   If more than two bits have to be added, then the rules must be extended. In practice the maximum number of bits to be added together at any time is 3.

$0 + 0 + 0 = 0$ carry 0
$0 + 0 + 1 = 1$ carry 0
$0 + 1 + 1 = 0$ carry 1
$1 + 1 + 1 = 1$ carry 1

Example 2.3
Add together the following binary numbers:
(a) $1010 + 0110$ (b) $1000 + 1111$ (c) $10110 + 10111$

Solution
Decimal equivalents are given in brackets.

(a)   1010   (10)
       +
             0110   (6)
             —————
            10000   (16)

(b)   1000   (8)
       +
             1111   (15)
             —————
            10111   (23)

(c)   10110   (22)
       +
            10111   (23)
            ——————
           101101   (45)

Note that, in addition, the upper number is referred to the **augend**, the lower number as the **addend** and the total as the **sum**.

A practical problem emerges from these examples. If, for instance, the numbers 1010 and 0110 are to be added as in Example 2.3 (a) above, four-bit registers could be used to hold each of the values to be added but the result, 10000, would need a five-bit register. This is known as overflow, and cannot be allowed to occur if the original wordlength is to be maintained. If this is the case, then only numbers that give a total within the permitted wordlength can be added. This problem will be considered in greater detail later in the chapter as will the case of negative numbers.

Subtraction

The subtraction process also needs a set of rules:

$$0 - 0 = 0 \quad \text{borrow } 0$$
$$1 - 0 = 1 \quad \text{borrow } 0$$
$$0 - 1 = 1 \quad \text{borrow } 1$$
$$1 - 1 = 0 \quad \text{borrow } 0$$

The situation can occur where more than one number is to be subtracted. The rules need an extension to take this possibility into account.

$$0 - 0 - 0 = 0 \quad \text{borrow } 0$$
$$0 - 0 - 1 = 1 \quad \text{borrow } 1$$
$$0 - 1 - 1 = 0 \quad \text{borrow } 1$$
$$1 - 0 - 0 = 1 \quad \text{borrow } 0$$
$$1 - 0 - 1 = 0 \quad \text{borrow } 0$$
$$1 - 1 - 1 = 1 \quad \text{borrow } 1$$

Example 2.4
Perform the following binary substractions:
(a) $1111 - 0011$ (b) $1001 - 0110$ (c) $110010 - 001011$

Solution
Decimal equivalents are given in brackets

(a)   1111   (15)

          —

          0011   (3)
          ‾‾‾‾
          1100   (12)

(b)   1001   (9)

          —

          0110   (6)
          ‾‾‾‾
          0011   (3)

(c)   110010   (50)

       —

       001011   (11)
       ‾‾‾‾‾‾
       100111   (39)

Note that the upper number is referred to as the **minuend,** the lower number as the **substrahend** and the result as the **difference**.

Clearly these problems have been carefully chosen so that the minuend is always larger than the subtrahend, the possibility of obtaining a negative result has been suppressed. The problem of negative numbers and how they are dealt with in practical digital systems will now be considered.

2.8 Negative binary numbers

First of all it will be assumed that all numbers are to be four bits long, residing in four-bit registers. The number zero will be represented by 0000, and

    +1 will be 0001
    +2 will be 0010
    +3 will be 0011

and so on.

If +3 is represented by 0011, then it can be deduced that +2 is represented by 0010 by subtracting 1 from 0011:

          0011   (+3)

          —

          0001   (+1)
          ‾‾‾‾
          0010   (+2)

Similarly   to   find −1:

          0000   (+0)

          —

 borrow 1 0001   (+1)
          ‾‾‾‾
          1111   (−1)

A problem occurs here in that there is a borrow left over. However, as the wordlength is 4, ignoring the borrow will not matter. What is left is 1111 which, interpreted as a normal binary number, corresponds to 15. This is not surprising since, by subtracting 1 from 0000 *and* ignoring the borrow, 1 is effectively subtracted from 16 (i.e. 10000) not 1 from 0000. Note that with a wordlength of 3, say, then it would be ignoring a borrow from a value of 8, not 16, and would leave 7, not 15.

$$\begin{array}{r} 000 \quad (+0) \\ - \\ \text{borrow} \quad 1\ 001 \quad (+1) \\ \hline 111 \quad (-1) \end{array}$$

If the wordlength was 5, then the process would be

$$\begin{array}{r} -\ 00000 \quad (+0) \\ \text{borrow} \quad 1\ 00001 \quad (+1) \\ \hline 11111 \quad (-1) \end{array}$$

giving a result of 31, i.e. 1 less than 32.

Generally speaking, with a wordlength N, the value obtained for -1 using this method is $2^N - 1$. For instance if $N = 4$, $2^4 = 16$, $2^4 - 1 = 15$, and so on. Each value *does* represent -1 for the system wordlength, although the actual number, of 1s will be different for different values of N.

$$\begin{array}{ll} \text{If} & 1111 = -1 \\ \text{then} & 1111 - 1 \quad \text{should represent} \ -2 \end{array}$$

$$\begin{array}{r} 1111 \quad (-1) \\ -\ 0001 \quad (+1) \\ \hline 1110 \quad (-2) \end{array}$$

and so on to calculate -3, -4, etc.

This can be tabulated:

1000 (-8)	0000 $(+0)$
1001 (-7)	0001 $(+1)$
1010 (-6)	0010 $(+2)$
1011 (-5)	0011 $(+3)$
1100 (-4)	0100 $(+4)$
1101 (-3)	0101 $(+5)$
1110 (-2)	0110 $(+6)$
1111 (-1)	0111 $(+7)$

It can be seen that if zero is taken to be positive, all the numbers starting with 0 are positive, all the numbers starting with 1 are negative, and there are eight positive numbers (0 to 7) and eight negative numbers (-1 to -8). The MSB can be taken to be a **sign bit**. The sign bit can be detected by circuitry, and the number interpreted as positive or negative accordingly. If this is done the numbers are interpreted as **signed binary numbers**.

Of course if 1 is added to $+7$ a problem occurs.

$$\begin{array}{r} 0111 \quad (+7) \\ + \\ 0001 \quad (+1) \\ \hline 1000 \qquad (?) \quad \neq\ +8 \end{array}$$

Similarly if 1 is subtracted from -8.

$$1000 \quad (-8)$$
$$- \quad 0001 \quad (+1)$$
$$\overline{0111} \qquad (?) \neq -9$$

The implication of all this is that it is only possible to represent numbers between -8 and $+7$ with a wordlength of 4, i.e. the number n is given by

$$+7 \geqslant n \geqslant -8$$

or alternatively

$$2^3 - 1 \geqslant n \geqslant -2^3$$

or with a wordlength of N

$$2^{N-1} - 1 \geqslant n \geqslant -2^{N-1}$$

This is known as the **number range** for a given wordlength. For example with $N = 8$:

$$2^7 - 1 \geqslant n \geqslant -2^7$$
$$+127 \geqslant n \geqslant -128$$

So, with a wordlength of 8, the largest positive number that can be represented is 127 and the most negative -128.

But is this system suitable for use in arithmetic calculations? By rewriting part of the previous list, and interpreting the numbers in the conventional way:

$$
\begin{array}{lll}
(-3) & 1101 \longrightarrow 13 & (16-3) \\
(-2) & 1110 \longrightarrow 14 & (16-2) \\
(-1) & 1111 \longrightarrow 15 & (16-1) \\
(0) & 0000 \longrightarrow 0 & \\
(+1) & 0001 \longrightarrow 1 & \\
(+2) & 0010 \longrightarrow 2 & \\
(+3) & 0011 \longrightarrow 3 &
\end{array}
$$

It becomes evident that the numbers assumed to be negative numbers in signed binary from are the difference between the magnitude of the number concerned and 16 (2^4). So -4 would be $16 - 4 = 12$ in decimal from, or 1100 in binary form. However, if a wordlength of 5 had been assumed, then the negative values would be obtained by subtraction from 32 (2^5).

Generally, the negative equivalent of a number is obtained by subtracting the magnitude of the number from 2^N where N is the wordlength. For this reason the negative values are referred to as being represented in two's **complement** form.

Example 2.5
Calculate the two's complement of the following numbers, assuming a wordlength of N.
(a) 7 ($N = 4$)   (b) 25 ($N = 6$)   (c) 193 ($N = 10$)

Solution
(a) If $N = 4$, then 7 must be subtracted from 2^4, i.e. 16. Therefore

$$\text{two's complement of } 7 = 16 - 7 = 9$$

which equals 1001 in binary.

(b) If $N = 6$, then 25 must be subtracted from 2^6, i.e. 64. Therefore

two's complement of $25 = 64 - 25 = 39$

which equals 100111 in binary.

(c) If $N = 10$, then 193 must be subtracted from 2^{10}, i.e. 1024. Therefore

two's complement of $192 = 1024 - 193 = 831$

$$
\begin{array}{rll}
2 & 831 & \\
2 & 415 & \text{r } 1 \quad \text{LSB} \\
2 & 207 & \text{r } 1 \\
2 & 103 & \text{r } 1 \\
2 & 51 & \text{r } 1 \\
2 & 25 & \text{r } 1 \\
2 & 12 & \text{r } 1 \\
2 & 6 & \text{r } 0 \\
2 & 3 & \text{r } 0 \\
2 & 1 & \text{r } 1 \\
& 0 & \text{r } 1 \quad \text{MSB}
\end{array}
$$

which is equivalent to 1100111111 in binary.

2.9 Complement arithmetic

Having established how to determine the two's complement of a number, it is now necessary to test its suitability for arithmetic operations. The two's complement of 7 for a 4-bit word was calculated in Example 2.5 (a) to be 1001. If this corresponds to -7, then adding it to $+7$ should result in the answer zero.

$$
\begin{array}{rl}
(-7) & 1001 \\
+ & \\
(+7) & 0111 \\
\hline
\text{Carry} & 0000
\end{array}
$$

Adding 9 (1001) to 7 (0111) as unsigned binary integers gives a result of 16 (10000). The 1 in the MSB corresponds to the ignored carry when $+7$ and -7 were added together above. As the two's complement of a four-bit number, n, is defined as $16 - n$, then this will occur in all cases, not only for four-bit numbers but for any wordlength. Thus adding a number to its complement will always give zero if the final carry is ignored. This is a prime requirement for a system of number representation.

A quicker method of obtaining the two's complement of a number is to invert all the bits and add 1. This system works in all cases for any wordlength. For instance, in Example 2.5 (a):

$7 = 0111$ (4 bits)
reversing the bits gives 1000
adding 1 gives 1001

In 2.5 (b):

25 = 011001 (6 bits)
reversing the bits gives 100110
adding 1 gives 100111

In 2.5 (c):

193 = 0011000001 (10 bits)
reversing the bits = 1100111110
adding the bits = 1100111111

all of which correspond to the previous results.

The reversed number, which is always 1 less than the two's complement, is called the **one's complement** for that reason. This procedure of bit reversal and adding 1 makes it very easy to implement in circuit form, and is a widely used method of obtaining the two's complement. Circuit details follow later in this text.

The next step is to try some arithmetic using the two's complement approach.

Example 2.6

Perform the following subtraction using two's complement arithmetic assuming a wordlength of N.

(a) $12 - 3$ $(N = 5)$ (b) $3 - 12$ $(N = 5)$ (c) $97 - 22$ $(N = 8)$ (d) $22 - 97$ $(N = 8)$

Solution

(a) Minuend = $+12$ = 01100 in a 5-bit word.
    Subtrahend = $+3$ = 00011 in 5 bits.
    Two's complement of subtrahend = 11101 = -3
Add the minuend to the two's complement of the subtrahend:

```
        01100      minuend (+ 12)
        11101      two's complement of subtrahend (− 3)
      ─────────
    ▸ 1 01001      difference = + 9
        ↑
        SB = + ve
ignore
carry
```

The result is $+9$ as given by 01001, remembering that the 0 in the MSB position of the 5-bit result corresponds to the sign bit, which is positive.

(b) Minuend = $+3$ = 00011
    Subtrahend = $+12$ = 01100
    Two's complement of subtrahend = 10100 = -12

Add the minuend to the two's complement of the subtrahend:

```
        00011      minuend (+ 3)
        10100      two's complement of subtrahend (− 12)
      ─────────
    0   10111      difference = − 9
        ↑
        SB = − ve
```

Note that this time the sign bit is negative. This means that the result obtained is in two's complement form.

A computer using this system would be just as happy with this result as with a positive. To check its value however it needs to be complemented.

result = 10111

which means of course that the answer is negative.

two's complement of 10111
is 01001 = 9

The result is −9 as expected.

(c) Minuend = +97 = 01100001 in 8 bits
 Subtrahend = +22 = 00010110 in 8 bits
 Two's complement of subtrahend = 11101010 (−22)

Add the minuend to the two's complement of the subtrahend:

```
        01100001    minuend (+ 97)
        11101010    two's complement of subtrahend
```

     ┌──→1 01001011
ignore   ↑   result is + 75
carry   SB = + ve

(d) Minuend = +22 = 00010110
 Subtrahend = +97 = 01100001
 Two's complement of subtrahend = 10011111 (−97)
 Add the minuend to the two's complement of the subtrahend:

```
        00010110    minuend (+ 22)
        10011111    two's complement of subtrahend (− 97)
```

     0 10110101
        ↑
        SB − ve   result is in two's complement form

Therefore complement result→01001011 = 75

answer is − 75

It is worth investigating one of these problems to see what is happening.
 Taking Example 2.6 (a):

12 − 3 = 9   was the effective calculation.

The actual calculation was:

12 + (−3) = 9 because of the complement approach.

or, more accurately,

12 + (32 − 3) − 32 = 9
 = 12 + 29 − 32 = 9
 = 01100 + 11101 − 100000 = 01001

two's complement   disregarded carry

So in fact there was no trickery involved at all.

2.10 Multiplication of binary numbers

Having dealt with addition and subtraction of binary numbers, it is now time to consider multiplication. Multiplication in binary can be carried out by the long multiplication process that was used before the advent of calculators, if slide rules and log tables were not available.

Example 2.7
Multiply the numbers 0101 and 0011 by the long multiplication process.

Solution

Bit number	3210	
	0101	Multiplicand (MD)
	0011	Multiplier (MR)
	0101	Bit 0 of MR = 1 so write down MD
Partial	0101–	Bit 1 of MR = 1 so write down MD left shifted
products	0000––	Bit 2 of MR = 0 so write down 0's left shifted
	0000–––	Bit 3 of MR = 0 so write down 0's left shifted
	00001111	Total of partial products = 15 = 5 × 3

The procedure outlined above gives the correct result, but is difficult to implement in circuit form as all the partial products have to be added together simultaneously at the final stage and have to be stored separately prior to this. With large wordlengths, the number of partial products is also large and so partial product storage becomes a serious problem. This problem can be simply alleviated by using a single register (hardware device for storing a word) to accumulate the partial products, by immediately adding the multiplicand when required and shifting the register contents to the right between additions. The register that holds the accumulating partial products is called the **accumulator**. This means that it is only ever necessary to add two binary numbers together at any one time, which considerably simplifies the circuitry required to perform the addition.

In order to utilise fully the wordlength number range of the multiplier and multiplicand, it is usual to have a double-length accumulator. For example if the wordlength was 2, ignoring sign bits for the present;

$$3_{10} = 11_2$$

If the multiplicand $= 11_2$ and the multiplier $= 11_2$ then the product is

$$3_{10} \times 3_{10} = 9_{10} = 1001_2$$

which requires four bits, i.e. double the original wordlength. This applies generally and is the reasoning behind the **double-length accumulator**.

Example 2.8
Multiply 12 by 9 assuming a wordlength of 8.

Solution

$$\text{Multiplicand} = 12_{10} = 00001100_2$$
$$\text{Multiplier} = 9_{10} = 00001001_2$$

Remember that MD means multiplicand and MR means multiplier.

	Multiplier	Accumulator content		Comment
		00000000	00000000	Initially set to zero
Multiplier bit (0) (LSB)	1	00001100	00000000	MR = 1 so add MD
		00001100	00000000	Partial product
		00000110	00000000	Shift right
bit 1	0	00000110	00000000	MR = 0 so add zero
		00000011	00000000	Shift right
bit 2	0	00000011	00000000	MR = 0 so add zero
		00000001	10000000	Shift right
bit 3	1	00001100	00000000	Add MD
		00001101	10000000	Partial product
		00000110	11000000	Shift right
bit 4	0	00000011	01100000	Add zero and shift right
bit 5	0	00000001	10110000	Add zero and shift right
bit 6	0	00000000	11011000	Add zero and shift right
bit 7	0	00000000	01101100	Add zero and shift right **Stop**

Result $= 0000000001101100_2$
which is 108_{10}, as expected.

Example 2.9
Multiply -17 by 24, assuming a wordlength of 8.

Solution

Multiplier $= 00011000_2$
Multiplicand $= -17_{10}$

It is necessary to find the two's complement of the multiplicand magnitude:

$+17_{10} = 00010001_2$

Two's complement $= 11101111_2 = 17_{10} =$ multiplicand.
Using the same procedure as in Example 2.8:

Multiplier = 00011000	Multiplier	Accumulator content		Comment
		00000000	00000000	Initially set to zero
Multiplier bit 0 (LSB)	0	00000000	00000000	Add zero and shift right
bit 1	0	00000000	00000000	Add zero and shift right
bit 2	0	00000000	00000000	Add zero and shift right
bit 3	1	11101111	00000000	Add MD
		11101111	00000000	Partial product
		11110111	10000000	Shift right
bit 4	1	11101111	00000000	Add MD
		11100110	10000000	Partial product
		11110011	01000000	Shift right
bit 5	0	11111001	10100000	Add zero and shift right
bit 6	0	11111100	11010000	Add zero and shift right
bit 7	0	11111110	01101000	Add zero and shift right

result = 1111111001101000_2
complementing gives 0000001100011000_2
which is 408_{10}
The product is therefore -408

When the shift right takes place, the sign bit is regenerated; i.e. if the sign bit was 0 then it is replaced by 0, while if it was 1 then it is replaced by 1. This may go unnoticed if the sign bit was 0, but looks a little strange if the sign bit was 1. Most computers and microprocessors have a facility for shifting numbers with sign bit regeneration. This type of shift is known as a arithmethic shift. Arithmetic shifts are essential if signed binary numbers are right shifted, otherwise a negative number would change to positive.

2.11   Division of binary numbers

Division is the reverse process to multiplication. In multiplication, the end result was a double length product. In division it is necessary to start with a double length dividend. The divisor will be single length to give a single length quotient.

Example 2.10

Divide 29 by 4 using the long division method. Assume a wordlength of 8.

Solution

In the working which follows, DD means dividend.

$$\text{Divisor} = 00000100_2$$

Quotient

Dividend	0 0 0 0 0 0 0 0 0 0 0 1 1 1 0 1		
		Quotient	bit number Q_n
Subtract divisor	0 0 0 0 0 1 0 0		
Sign bit ⟶	1 1 1 1 1 1 0 0		
= 1 so restore the			
dividend by	0 0 0 0 0 1 0 0		
adding divisor	0 0 0 0 0 0 0 0 0		
Right shift and			
subtract divisor	0 0 0 0 0 1 0 0		
SB = 1 ⟶	1 1 1 1 1 1 0 0	0	Q_7
$Q_7 = 0$, restore DD	0 0 0 0 0 1 0 0		
	0 0 0 0 0 0 0 0 0		
Right shift and			
subtract divisor	0 0 0 0 0 1 0 0		
SB = 1 ⟶	1 1 1 1 1 1 0 0	0	Q_6
$Q_6 = 0$, restore DD	0 0 0 0 0 1 0 0		
	0 0 0 0 0 0 0 0 0		
Right shift and			
subtract divisor	0 0 0 0 0 1 0 0		
SB = 1 ⟶	1 1 1 1 1 1 0 0	0	Q_5
$Q_5 = 0$, restore DD	0 0 0 0 0 1 0 0		
	0 0 0 0 0 0 0 0 1		
Right shift and			
subtract divisor	0 0 0 0 0 1 0 0		
SB = 1 ⟶	1 1 1 1 1 1 0 1	0	Q_4
$Q_4 = 0$, restore DD	0 0 0 0 0 1 0 0		
	0 0 0 0 0 0 1 1		
Right shift and			
subtract divisor	0 0 0 0 0 1 0 0		
SB = 1 ⟶	1 1 1 1 1 1 1 1	0	Q_3
$Q_3 = 0$, restore DD	0 0 0 0 0 1 0 0		
	0 0 0 0 0 1 1 1		
Right shift and			
subtract divisor	0 0 0 0 0 1 0 0		
SB = 0, $Q_2 = 1$	0 0 0 0 0 0 1 1 0	1	Q_2
Right shift and			
subtract divison	0 0 0 0 0 1 0 0		
SB = 0, $Q_1 = 1$	0 0 0 0 0 0 1 0 1	1	Q_2
Right shift and			
subtract divisor	0 0 0 0 0 1 0 0		
SB = 0, $Q_0 = 1$	0 0 0 0 0 0 0 1	1	Q_0

Stop

So quotient = 00000111 (+ 7)

and remainder = 00000001 (+ 1)

The routine adopted is as follows.

(a) Subtract the divisor from the most significant byte of the dividend. If the result is positive then the quotient will be greater than eight bits long and so this approach will not work.
(b) If the result is negative then the dividend should be restored by adding the divisor.
(c) Right shift and subtract the divisor.
    If the result is negative, set Q_n to 0 and restore the dividend. If $n = 0$ then stop. If $n > 0$ then go back to (c) above. If the result is positive, set Q_n to 1. If $n = 0$ then stop. If $n > 0$ then go back to (c) above.

In this example, when Q_n became equal to Q_0 a remainder of 00000001 was left. This will clearly only be zero if the numbers divide exactly.

This method of approach could be implemented by hardware, i.e. a purpose-built circuit to follow the above algorithm, or by software, using the appropriate sequence of operation codes to subtract, shift, etc, where necessary.

2.12 The octal number system

The preceding sections have shown that binary calculations can be fairly complex, leading to some confusion with a large number of 0s and 1s. One method of abbreviating large binary numbers is to convert them to octal. A binary number is three times longer, i.e. it contains three times the number of digits than its octal equivalent. A great number of computer systems use the octal system by incorporating an octal keyboard or keypad. In the decimal system there are ten symbols 0–9 as seen earlier in this chapter. The octal system has a radix of 8, and so there are eight symbols 0–7.

$$0 \; 1 \; 2 \; 3 \; 4 \; 5 \; 6 \; 7$$

There is no symbol corresponding to the radix. As in all number systems, the symbols corresponding to the radix are 10. The weighting sequence is therefore.

$$\text{Octal point}$$
$$\ldots 8^2 \, 8^1 \, 8^0 \cdot 8^{-1} \, 8^{-2} \, 8^{-3} \ldots$$
$$\underset{\text{integers}}{\longleftarrow} \quad \underset{\text{fractions}}{\longrightarrow}$$

The octal system will follow the number sequence:

	0	1	2	3	4	5	6	7	← Run out of symbols so
First Repeat	10	11	12	13	14	15	16	17	reuse them with a prefixed repeat number.
Second Repeat	20	21	22	23	24	25	26	27	
Third Repeat	30	31	32	. . . etc.					

An octal number can be recognised by the '8' subscript, for instance

$$275_8$$

Decimal and octal equivalents up to 16_{10} are listed below:

Decimal	Octal	Decimal	Octal
0	0	9	11
1	1	10	12
2	2	11	13
3	3	12	14
4	4	13	15
5	5	14	16
6	6	15	17
7	7	16	20
8	10		

Note that $8_{10} = 10_8$
$16_{10} = 20_8$
Similarly $24_{10} = 30_8$
$32_{10} = 40_8$

It is useful to know the following powers of 8:

$8^0 = \qquad 1 = 2^0$
$8^1 = \qquad 8 = 2^3$
$8^2 = \qquad 64 = 2^6$
$8^3 = \qquad 512 = 2^9$
$8^4 = \quad 4096 = 2^{12}$
$8^5 = 32768 = 2^{15}$

It is no coincidence that powers of 8 correspond to exact powers of 2. It was for that reason the octal number system was chosen, as was pointed out earlier in this chapter.

Octal to decimal conversion is fairly straightforward, as the following example will show.

Example 2.11
Convert the following octal numbers to decimal:
(a) 7_8 (b) 13_8 (c) 265_8

Solution
(a)   Octal numbers up to and including 7_8 are the same as decimal, so $7_8 = 7_{10}$.
(b) $13_8 = (1 \times 8^1) + (3 \times 8^0)$
$= 8 + 3$
$= 11_{10}$
(c) $265_8 = (2 \times 8^2) + (6 \times 8^1) + (5 \times 8^0)$
$= (2 \times 64) + (6 \times 8) + (5 \times 1)$
$= 128 + 48 + 5$
$= 181_{10}$

When converting into the binary number system (which has a radix of 2) from decimal, the decimal number is repeatedly dividend by 2. To convert from decimal to octal, the decimal number is repeatedly dividend by 8. In each case it is the remainders that provide the required conversion. This is illustrated in the following example:

Example 2.12
Convert the decimal number 150_{10} into octal.
Solution

```
8 150

8  18     r 6   LSD

8   2     r 2

    0     r 2   MSD
```

So $150_{10} = 226_8$

Consider the following bit pattern:

000000001_2

Multiplying by 2 gives:

000000010_2

Multiplying by 2 again gives:

000000100_2

and by two again gives:

000001000_2

which of course corresponds to multiplying the original number by 8. Shifting a 1 three places to the left is equivalent to multiplying by 8, therefore, and a link with the octal system can be established. In octal, shifting *one* place to the left is equivalent to multiplication by 8. So three binary digits are equivalent to one octal digit. If a bit pattern is broken down into groups of three bits, the situation becomes clearer. Taking the bit pattern 000101110_2 and putting it into groups of three gives:

000 101 110

Remembering that octal numbers up to 7_8 are equivalent to decimal, the octal equivalent can be derived.

$$000 \quad 101 \quad 110_2$$
$$\searrow \quad \downarrow \quad \swarrow$$
$$0 \quad 5 \quad 6$$

So $000101110_2 = 056_8$

This can be easily checked by converting each value to decimal:

$$000101110_2 = 32 + 8 + 4 + 2 = 46_{10}$$
$$056_8 = (0 \times 8^2) + (5 \times 8^1) + (6 \times 8^0)$$
$$= 46_{10}$$

The bit pattern chosen had a wordlength of 9, which is exactly divisible by 3. This will not always be the case. For example, the 16-bit word:

$$1101101101110110$$
$$= 1 \ 101 \ 101 \ 101 \ 110 \ 110_2$$

in octal: 1   5   5   5   6 6_8

It is important to avoid incorrect grouping. With integers the best policy is to add leading zeros to the number to get complete groups of 3. Fractions or mixed numbers can be treated in the same way, for example

$$01110101.10101110_2$$
$$= 01\ 110\ 101\ .\ 101\ 011\ 10_2$$

Note that the groups are formed by working from the binary point outwards. Zeros can be added to the outer groups to make them complete:

$$001\ 110\ 101\ .\ 101\ 011\ 100$$

which is $\quad 1\quad 6\quad 5\ .\ 5\quad 3\quad 4_8$

Example 2.13
(a) Convert to octal: (i) 0101101_2 (ii) 1011101101101_2 (iii) 10110.0111_2
(b) Convert to binary: (i) 375_8 (ii) 101_8 (iii) 473.201_8

Solution
(a)   (i) 0101101_2 $\qquad = 000\ 101\ 101_2$
$\qquad\qquad\qquad\qquad\qquad = 0\quad 5\quad 5_8$
Thus $\qquad 0101101_2 \qquad = 055_8$
$\qquad$ (ii) $1011101101101_2 = 001\ 011\ 101\ 101\ 101_2$
$\qquad\qquad\qquad\qquad\qquad = 1\quad 3\quad 5\quad 5\quad 5_8$
Thus $\qquad 1011101101101_2 = 13555_8$
$\qquad$ (iii) $10110.0111_2 \qquad = 010\ 110\ .\ 011\ 100_2$
$\qquad\qquad\qquad\qquad\qquad = 2\quad 6\ .\ 3\quad 4_8$
Thus $\qquad 10110.0111_2 \qquad = 26.34_8$
(b)   (i) $375_8 \qquad\qquad\qquad = 011\ 111\ 101_2$
$\qquad$ (ii) $101_8 \qquad\qquad\qquad = 001\ 000\ 001_2$
$\qquad$ (iii) $473.201_8 \qquad\quad = 100\ 111\ 011\ .\ 010\ 000\ 001_2$

It is more efficient to perform an intermediate octal conversion when converting a large decimal number to binary. The decimal to octal conversion is more rapid and the octal to binary conversion is simple. The following example will illustrate this.

Example 2.14
Convert 1000_{10} to binary (a) by the conventional method and (b) by an intermediate octal conversion.

Solution

(a)

2	1000		
2	500	r 0	LSB
2	250	r 0	
2	125	r 0	
2	62	r 1	
2	31	r 0	
2	15	r 1	
2	7	r 1	
2	3	r 1	
2	1	r 1	
	0	r 1	MSB

So $1000_{10} = 1111101000_2$

(b)

8	1000		
8	125	r 0	LSD
8	15	r 5	
8	1	r 7	
	0	r 1	MSD

which gives $1750_8 = 001111101000_2$

Clearly method (b) is more efficient

2.13 The hexadecimal number system

Although the octal number system is still widely used, a more commonly used system nowadays is hexadecimal (hex). Microprocessors particularly use hex as a shorthand for binary and a knowledge of it is essential. In the hexadecimal system the radix is 16 and so there are 16 symbols. Ten of these symbols (0–9) can be stolen from the decimal system. Another six are needed, however, and the convention now is to use the letters of the alphabet A–F for these. The hexadecimal symbols are tabulated below together with their decimal equivalents.

Decimal	Hexadecimal	Decimal	Hexadecimal
0	0	8	8
1	1	9	9
2	2	10	A
3	3	11	B
4	4	12	C
5	5	13	D
6	6	14	E
7	7	15	F

The weighting sequence is:

$$\ldots 16^3 \ 16^2 \ 16^1 \ 16^0 \ . \ 16^{-1} \ 16^{-2} \ 16^{-3} \ 16^{-4} \ldots$$

$$\underleftarrow{\qquad} \ \underrightarrow{\qquad}$$
$$\text{integers} \quad \text{fractions}$$

The number sequence is:

0 1 2 3 4 5 6 7 8 9 A B C D E F
10 11 12 13 14 15 16 17 18 19 1A 1B 1C 1D 1E 1F
20 21 22 23 24 25 26 27 28 29 2A . . . etc.

The subscript 16 or H is usually added to indicate a hexadecimal number. For example:

$$3A7_{16} \text{ or } 3A7_H$$

Note that $16_{10} = 10_{16}$ or 10_H
$32_{10} = 20_{16}$ or 20_H
$48_{10} = 30_{16}$ or 30_H

and so on

With a radix of 16, the value of the weightings increase very rapidly:

$16^0 = 1 \qquad = 2^0$
$16^1 = 16 \qquad = 2^4$
$16^2 = 256 \quad = 2^8$
$16^3 = 4096 = 2^{12}$ etc.

With octal numbers, great use is made of the fact that 8 was a power of 2. Hexadecimal is used for the same reason – i.e. 16 is a power of 2.

Hexadecimal to decimal conversion is not difficult. The next example will illustrate this.

Example 2.15
Convert the following hexadecimal numbers to decimal:
(a) C_{16} (b) $4D_{16}$ (c) $FADE_{16}$

Solution

(a) C_{16} $= 12_{10}$

(b) $4D_{16}$ $= (4 \times 16^1) + (D \times 16^0)$

                    $= 64 + 13$

                    $= 77_{10}$

(c) $FADE_{16} = (F \times 16^3) + (A + 16^2) + (D \times 16^1) + (E \times 16^0)$

                    $= (15 \times 4096) + (10 \times 256) \times (13 \times 16) + (14 \times 1)$

                    $= 61440 + 2560 + 208 + 14$

                    $= 64222_{10}$

Decimal to hex conversion is achieved by successive division by 16.

Example 2.16

Convert the following decimal numbers into hexadecimal:

(a) 25_{10} (b) 200_{10} (c) 452_{10}

Solution

(a)   16 | 25

            16 | 1   r 9   LSD

               | 0   r 1   MSD

So $25_{10} = 19_{16}$

(b)   16 | 200

            16 | 12   r 8   LSD

               | 0   r C   MSD

So $200_{10} = C8_{16}$

(c)   16 | 452

            16 | 28   r 4   LSD

            16 | 1   r C

               | 0   r 1   MSD

So $452_{10} = 1C4_{16}$

In the same way that bit patterns were split into groups of three for every conversion into octal, they can be split into groups of four for conversion into hexadecimal. For instance the 16-bit word:

$$1101101101110110_2$$
$$= 1101\ 1011\ 0111\ 0110_2$$
$$= \quad D \quad B \quad 7 \quad 6_{16} \qquad \text{in hexadecimal}$$

This works because $16 = 2^4$, so shifting left by four binary digits is equivalent to shifting left by 1 hex digit.

Example 2.17

(a) Convert to hexadecimal: (i) 0101101_2 (ii) 10111101101101_2

(b) Convert to binary: (i) $DEAF_{16}$ (ii) $1CED_{16}$

Solution
(a)   (i) 0101101_2 $\qquad = 0010\ 1101_2$
$\qquad\qquad\qquad\qquad\qquad = 2D_{16}$
Thus $0101101_2 \qquad = 2D_{16}$
$\qquad$ (ii) $10111101101101_2 \ = 0010\ 1111\ 0110\ 1101_2$
$\qquad\qquad\qquad\qquad\qquad = 2 \qquad F \quad 6 \qquad D_{16}$
Thus $10111101101101_2 \quad = 2F6D_{16}$
(b)   (i) $DEAF_{16} = 1101\ 1110\ 1010\ 1111_2$
$\qquad$ (ii) $ICED_{16} = 0001\ 1100\ 1110\ 1101_2$

Although the hexadecimal number system can represent fractions and mixed numbers, in most cases it is used only for integer work. A common application is to specify memory addresses in mircroprocessor systems. For instance, a microprocessor memory system with 64K ($65\ 536_{10} = 2^{16}$) of memory would have hexadecimal addresses running from 0000_{16} to $FFFF_{16}$ ($65\ 535_{10}$).

2.14 Binary coded decimal numbers

Although it may be convenient to use binary numbers in preference to decimal numbers in digital circuits, in most cases the display of numbers is required to be in decimal format. It has been seen that the use of hexadecimal and octal is convenient because of their close relationship to binary and that decimal numbers do not have this relationship with binary numbers. By the use of a special code, a similar relationship between decimal numbers and a four-bit pattern can be established. In this way each decimal digit can be converted directly into the corresponding four-bit pattern in binary coded decimal (BCD) format. The selection of bit patterns to correspond to decimal digits is arbitrary and usually depends upon the particular application. Some examples of these important codes are given in Chapter 12.

2.15 Problems

2.1 Convert to binary:
(a) 17_{10} (b) 96_{10} (c) 255_{10} (d) 1025_{10}

2.2 Convert to binary:
(a) 0.5_{10} (b) 0.05_{10} (c) 0.975_{10} (d) 0.83_{10}

2.3 Convert to binary:
(a) 26.92_{10} (b) 183.02_{10} (c) 596.103_{10} (d) 20.0003_{10}

2.4 Add together the following binary numbers:
(a) $1101_2 + 0010_2$ (b) $11111_2 + 10110_2$ (c) $1101_2 + 10111_2$ (d) $010101_2 + 0011010_2$

2.5 Using normal subtraction techniques perform the following:
(a) $11110_2 - 01011_2$ (b) $1011_2 - 0111_2$ (c) $11111_2 - 10100_2$
(d) $0110110_2 - 0001001_2$

2.6 What is the number range of the following wordlengths for signed binary integers?
(a) 3   (b) 7   (c) 10   (d) 16

2.7 Calculate the two's complement in binary form for the following numbers, assuming a wordlength of N.
(a) 13 ($N = 5$)   (b) 27 ($N = 7$)   (c) 120 ($N = 16$)   (d) 1023 ($N = 16$)

2.8 Perform the following subtractions using two's complement arithmetic, assuming a wordlength of N.

(a) $13 - 6$ ($N = 6$)   (b) $6 - 13$ ($N = 6$)   (c) $125 - 62$ ($N = 16$)
(d) $62 - 125$ ($N = 16$)

2.9 Multiply the following numbers in binary, showing all working. Use the method that is easier to implement in circuit form.
(a) $27_{10} \times 12_{10}$ (wordlength $= 8$)   (b) $-5_{10} \times 15_{10}$ (wordlength $= 8$)

2.10 Divide 35 by 8 using the restoring method. Assume a wordlength of 8.

2.11 Convert the following octal numbers to decimal:
(a) 5_8 (b) 17_8 (c) 4000_8 (d) 15300_8
2.12 Convert the following decimal numbers into octal:
(a) 25_{10} (b) 376_{10} (c) 1023_{10} (d) 4096_{10}

2.13 Convert to octal:
(a) 01101_2 (b) 101101111001_2 (c) 101.0110111_2 (d) 11011.1110111101_2

2.14 Convert the following decimal numbers to binary by means of an intermediate octal conversion:
(a) 512_{10} (b) 4097_{10} (c) $16,300_{10}$ (d) $31,767_{10}$

2.15 Convert the following hexadecimal numbers to decimal:
(a) $D3_{16}$ (b) $F1C_{16}$ (c) $2A93_{16}$ (d) $F1F1_{16}$

2.16 Convert the following decimal numbers into hexadecimal:
(a) 123_{10} (b)972_{10} (c) 2163_{10} (d) 8000_{10}

2.17 Convert to hexadecimal:
(a) 1010111_2 (b) 1111011111100001_2 (c) 11010.101101_2 (d) 01011101.011011011101_2

2.18 Convert to binary:
(a) $D3_{16}$ (b) $F1C_{16}$ (c) $2A93_{16}$ (d) $F1F1_{16}$

3

LOGIC ALGEBRA

3.1 Introduction

Chapter 2 gives a fairly thorough treatment of the way in which bit patterns are used to represent numbers, and how they can be used in the arithmetic operations of add, subtract, multiply and divide. The idea of representing binary numbers by voltages signifying 0 and 1 logic levels has also been introduced. The circuit hardware that is required to manipulate the bit patterns will not be considered until Chapter 4. However, circuit operations can be defined in an abstract way by considering circuit outputs as logical functions of circuit inputs. In this way it is possible to design circuits to fulfil a logical function before attempting to implement it with hardware. To represent circuit operation, a form of algebra has been developed from the work initiated over a hundred years ago by George Boole. Boole was interested in the manipulation of logical propositions which could have only two possible answers, true or false. In the late thirties CE Shannon introduced Boole's method to the analysis and simplification of telephone switching circuits. Later, **Boolean** or **logic algebra** was adopted for digital computer work, and it now plays a very important part in the design and simplification of digital and microprocessor-based systems. It can be seen therefore that logic algebra is used to represent digital circuits involving variables which can have one of only two possible states, corresponding to the binary numbers 0 and 1. This algebra facilities the simplification of complex logic functions, and in turn reduces the total design cost due to the employment of fewer components. Consequently the reliability of the system is also improved. But first the rules of the logical processes involved will be examined.

3.2 Logical operations

The algebraic variables, often called **literals**, are usually denoted by letters of the alphabet such as A, B, C through to X, Y, Z. Any such variable may assume one of the two values 0 or 1, i.e. zero or one, representing an open switch or a closed switch, a low voltage state or a high voltage state, an open circuit connection or a short circuit connection. Since logic algebra deals with two-state variables, it is sometimes referred to as switching algebra. The variables can be operated on, or they may be combined to form **terms** or **functions**.

It must be stressed that the rules for solution, simplification or minimisation are not the same as for ordinary algebra. There are only three basic logical operations, AND, OR and NOT, which are defined as follows.

AND means that both variables (or terms) are present, and is denoted by the symbol. (a dot).

OR means that one or more variables (or terms) are present, and is denoted by the symbol + (a plus).

NOT indicates negation, inversion or complementation, and is denoted by the symbol ▬ (a bar) over the variable, term or binary number.

The three operations will be now discussed, with more detail, starting with the NOT function.

3.3   The NOT operation

If a variable is denoted by A then the inverse A is denoted by $\overline{A}$, which is read 'NOT A'.
Since the possible values of A are a 0 or a 1, if A = 1 then $\overline{A}$ = 0.
Similarly, $\overline{0}$ = 1 and $\overline{1}$ = 0 (read '*not* zero is one' and '*not* one is zero').
When logic algebra is applied to a network,

(i) an **open circuit** is denoted by a 0 or ▬▲ ●▬ , and
(ii) a **short circuit** is denoted by a 1 or ▬●▬ .

It will be shown later that most functions can be represented electrically as a combination of switches.

Note that complementation or inversion can be repeated several times. For instance

$$\overline{\overline{A}} = A \tag{3.1}$$

and $$\overline{\overline{\overline{B}}} = \overline{B} \tag{3.2}$$

or $$\overline{\overline{1}} = 1 \quad \text{and} \quad \overline{\overline{\overline{0}}} = \overline{\overline{1}} = \overline{0} = 1$$

showing that double inversion restores the original value. This is usually presented in a form of a **truth table**, which is a more convenient way of indicating whether the value of the function is true, i.e. a 1, or false, i.e. a 0. The 'NOT' truth table for the two possible values of A is shown in Table 3.1, where A is a variable or an input and F is a function or an output represented by the variable.

Table 3.1
Truth table
for NOT
operation

A	F
0	1
1	0

3.4 The AND operation

The AND operation, often referred to as the **logical product** or as **Boolean multiplication**, will now be considered. The AND operation is denoted by a dot, i.e. A.B, which is read 'A *and* B'.

If F denotes a function, then

$$F = A.B \qquad\qquad (3.3)$$
or simply $\qquad F = AB \qquad\qquad (3.4)$

where the dot indicating the AND operation is omitted.

With two variables, each represented by a binary 0 or 1, there can be $2^2 = 4$ different solutions for the function F.

The rules for AND operation on two binary numbers are as follows:

$0.0 = 0$
$0.1 = 0$
$1.0 = 0$
$1.1 = 1$

These show that the result is a 1 only if both digits are a 1.

The truth table for the AND function can now be written as shown in Table 3.2.

Table 3.2   Truth table for the AND operation of Equation 3.4

B	A	F
0	0	0
0	1	0
1	0	0
1	1	1

In general, the number of solutions for a function depends on the number of the variables, and is given by 2^n, where n is the number or variables.

Example 3.1
Derive the logical expression and the truth table for a three-variable AND function.

Solution
If the three variables are A, B and C, and the resulting AND function is represented by F, then

$$F = ABC \qquad\qquad (3.5)$$

which is read 'F equals A *and* B *and* C'.

The truth table for the three-variable function is given in Table 3.3.

Table 3.3   Truth table for AND operation given in Equation 3.5

C	B	A	F
0	0	0	0
0	0	1	0
0	1	0	0
0	1	1	0
1	0	0	0
1	0	1	0
1	1	0	0
1	1	1	1

Note that the function F is a 1 only if A, B and C are all 1.

In writing the truth table, the binary number system is used for determining all posible combinations of the variables which are forming the function F. In general, there is a choice of taking A as the least significant digit or as the most significant digit. This time A is taken as the least significant digit.

3.5 The OR operation

The logical OR operation is denoted by the + sign. This must not be confused with the ordinary algebraic plus sign. As a logical function operator the + sign means that, for two variables, A and B, the function F is given by

$$F = A + B \qquad (3.6)$$

which is read 'F equals A *or* B'. Thus the result of the operation is as follows: F is a 1 if A *or* B, or *both* A *and* B are 1. F is a 0 only if *both* A *and* B are 0. It must be emphasised here that the + sign is always be read 'OR', not 'plus' or 'and'.

The rules of the logical OR operation for a function with two variables are as follows:

$$0 + 0 = 0$$
$$0 + 1 = 1$$
$$1 + 0 = 1$$
$$1 + 1 = 1$$

The OR function can be used with more than two variables. The truth tables for two and three variables are shown in Tables 3.4 and 3.5.

Table 3.4   Truth table for OR operation given in Equation 3.6

B	A	F
0	0	0
0	1	1
1	0	1
1	1	1

Table 3.5   Truth table for OR operation of three variables

C	B	A	F
0	0	0	0
0	0	1	1
0	1	0	1
0	1	1	1
1	0	0	1
1	0	1	1
1	1	0	1
1	1	1	1

3.6 Other combinational operators

Three more logical functions are obtained as a result of the combinations of the basic three operations already discussed. The new logical operations are as follows.

(i)   The NAND operation. This is derived from the NOT AND combination, i.e. by inverting the AND function:

$$F = \overline{AB} \tag{3.7}$$

(ii)   The NOR operation. This is derived from the NOT OR combination, i.e. by inverting the OR function:

$$F = \overline{A + B} \tag{3.8}$$

(iii)   The exclusive or operation. The XOR, as it is usually referred to, is the AND OR AND combination of two variables which never are the same, giving

$$F = \overline{A}B + A\overline{B} \tag{3.9}$$

Very often the Equation 3.9 is written using the symbolic representation

$$F = A \oplus B \tag{3.10}$$

where the $\oplus$ sign is used to indicate the XOR operation.

   The truth tables for the NAND, NOR and XOR operations are given in Tables 3.6, 3.7 and 3.8 respectively.

Table 3.6   Truth table for NAND operation

A	B	AB	$F = \overline{AB}$
0	0	0	1
0	1	0	1
1	0	0	1
1	1	1	0

Table 3.7   Truth table for NOR operation

A	B	A + B	$F = \overline{A + B}$
0	0	0	1
0	1	1	0
1	0	1	0
1	1	1	0

Table 3.8   Truth table for XOR operation

A	B	$\overline{A}$	$\overline{B}$	$\overline{A}B$	$A\overline{B}$	$F = A \oplus B$
0	0	1	1	0	0	0
0	1	1	0	1	0	1
1	0	0	1	0	1	1
1	1	0	0	0	0	0

All the operations AND, NOR and XOR can be used with more than two variables, and will be demonstrated later.

3.7 Theorems and laws of logic algebra

The laws that govern logic or switching algebra will be discussed now by considering both the functional operation in symbolic form and an electrical representation using switches. The symbols used in this section and their physical realisation will be fully explained in Chapter 4.

AND Function

The AND function symbol is

Note that a dot equals AND, which is equivalent to a series connection.

$A.0 = 0$ (3.11)

$A.1 = A$ (3.12)

$A.A = A$ (3.13)

$A.\bar{A} = 0$ (3.14)

OR Function

Note that a + equals OR, which is equivalent to a parallel connection. The OR function symbol is

$A + B$

$A + 0 = A$ (3.15)

$A + 1 = 1$ (3.16)

$A + A = A$ (3.17)

$A + \overline{A} = 1$ (3.18)

If the Equations 3.14 and 3.18 are inverted, then

$$\overline{A.\overline{A}} = 1 = \overline{A} + A \tag{3.19}$$

$$\overline{A + \overline{A}} = 0 = \overline{A}.A \tag{3.20}$$

This leads to the following two **De Morgans theorems:**

$$\overline{A.B} = \overline{A} + \overline{B} \tag{3.21}$$

and

$$\overline{A + B} = \overline{A}.\overline{B} \tag{3.22}$$

The Equations 3.21 and 3.22 can be verified by writing truth tables, as shown in the following examples.

Example 3.2
Verify Equation 3.21 by writing the truth table.

Solution
In writing the truth table, a separate column is provided for each variable, for each subsequent operation on the variable and for the combination of the variables, as shown in Table 3.9.

Table 3.9   Truth table for Equation 3.21

A	B	AB	$\overline{AB}$	$\overline{A}$	$\overline{B}$	$\overline{A} + \overline{B}$
0	0	0	1	1	1	1
0	1	0	1	1	0	1
1	0	0	1	0	1	1
1	1	1	0	0	0	0

Thus the truth table reduces the complex task of evaluation of the function to a simple inversion of a 0 to a 1, or a 1 to a 0. It speeds up the solution and reduces the possibility of making errors, since only one operation is performed at a time.

In Table 3.9, column 4 is equal to column 7. Therefore the two sides of the Equation 3.21 are identical.

This method of verifying the equality is often referred to as **perfect induction.**

Example 3.3
Show that both sides of Equation 3.22 are identical.

Solution
The truth table is shown in Table 3.10, where each column is evaluated one by one.

Table 3.10   Truth table for Equation 3.22

A	B	A + B	$\overline{A + B}$	$\overline{A}$	$\overline{B}$	$\overline{A}, \overline{B}$
0	0	0	1	1	1	1
0	1	1	0	1	0	0
1	0	1	0	0	1	0
1	1	1	0	0	0	0

As in the previous example, columns 4 and 7 are equal, proving the validity of the expression.

Equations 3.21 and 3.22 can be extended to any number of variables. De Morgan's theorems therefore provide the means for transformation of logic expressions by conversion of AND statements to OR and vice versa.
  In general,

$$\overline{A.B.C \ldots} = \overline{A} + \overline{B} + \overline{C} + + + \tag{3.23}$$

which means that the complement of a product is equal to the sum of the complements.
  Similarly,

$$\overline{A + B + C + + +} = \overline{A}.\overline{B}.\overline{C} \ldots \tag{3.24}$$

which means that the complement of a sum is equal to the product of the complements.
  Therefore to obtain an inversion of a funtion, each variable must be inverted, each AND must be replaced by an OR sign, each OR sign must be replaced by an AND sign, and the whole expression thus obtained must also be inverted as shown in Example 3.4.

Example 3.4
Obtain an equivalent expression for (i) $\overline{A.B.C.D}$, and (ii) $\overline{A + B + C + D}$.

Solution
(i) Applying the general form of De Morgan's theorem, invert the varible, invert the sign, and invert the whole expression. This gives

$$\overline{A.B.C.D} = \overline{\overline{A} + \overline{B} + \overline{C} + \overline{D}} = \overline{A} + \overline{B} + \overline{C} + \overline{D} \tag{3.25}$$

since double inversion cancels.
(ii) Similarly,

$$\overline{A + B + C + D} = \overline{\overline{A}.\overline{B}.\overline{C}.\overline{D}} = \overline{A}.\overline{B}.\overline{C}.\overline{D} \tag{3.26}$$

There are two **commutative laws** relating to the order in which literals appear in a term or function. One is associated with an AND function (Equation 3.27), and the other with an OR function (Equation 3.28).

Commutative laws

$$A.B = B.A \qquad\qquad (3.27)$$

and

$$A + B = B + A \qquad\qquad (3.28)$$

These laws state that the order in which the variables or terms appear in the equation is irrelevant.

Associative Laws

The two **associative laws** which relate to the way in which literals may be grouped can take the form of an AND function (Equation 3.29) or an OR function (Equation 3.30).

$$ABC = (AB)C = A(BC) = (AC)B \qquad\qquad (3.29)$$

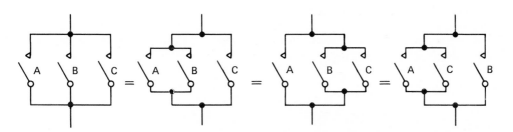

and

$$A + B + C = (A + B) + C = A + (B + C) = (A + C) + B \qquad\qquad (3.30)$$

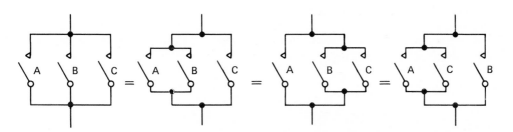

These laws state that the order in which the variables are bracketed is irrelevant. However,

$$(A + B)C + D \neq (A + B)(C + D) \qquad\qquad (3.31)$$

but

$$(A + B)C + D = C(A + B) + D = D + C(A + B) \qquad\qquad (3.32)$$

Distributive Laws

The **distributive laws** which define how literals may be distributed in a term or function are either in the form of Product of Sums. (Equation 3.33) or in the form of Sum of Products (Equation 3.34).

$$A(B + C) = AB + AC \tag{3.33}$$

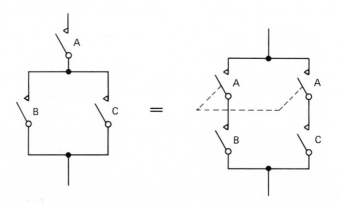

and

$$A + (BC) = (A + B)(A + C) \tag{3.34}$$

where switches A operate together.

To verify the above two equations, the rules of ordinary algebra are used, although the rules of logic algebra are different in many respects. In Equation 3.33, taking the right hand side, A is a common factor. Therefore

$$AB + AC = A(B + C) \tag{3.35}$$

Table 3.11 Truth table for Equation 3.33

A	B	C	AB	AC	AB + AC	B + C	A(B + C)
0	0	0	0	0	0	0	0
0	0	1	0	0	0	1	0
0	1	0	0	0	0	1	0
0	1	1	0	0	0	1	0
1	0	0	0	0	0	0	0
1	0	1	0	1	1	1	1
1	1	0	1	0	1	1	1
1	1	1	1	1	1	1	1

Otherwise the equation can be verified using the truth table method as shown in Table 3.11. Note that the columns 6 and 8 are identical. Therefore the operations A(B + C) and AB + AC will yield the same results for all possible values of the variables A, B and C.

Similarly, Equation 3.34 can be verified. Considering the right hand side of the equation gives

$$
\begin{aligned}
(A + B)(A + C) &= AA + AC + AB + BC \\
&= A \ \ + AC + AB + BC \\
&= A(1 + C + B) + BC \\
&= A + BC
\end{aligned}
$$ (3.36)

since

$$
\begin{aligned}
AA \qquad &= A \text{ (by Equation 3.13)} \\
1 + C + B &= 1 \text{ (by Equation 3.16)} \\
\text{and } A.1 \quad &= 1 \text{ (by Equation 3.12)}
\end{aligned}
$$

Example 3.5
Using the truth table method verify the Equation 3.34.

Solution
The truth table for the Equation 3.34 is shown in Table 3.12.

Table 3.12   Truth table for Equation 3.34

A	B	C	BC	A + BC	A + B	A + C	(A + B)(A + C)
0	0	0	0	0	0	0	0
0	0	1	0	0	0	1	0
0	1	0	0	0	1	0	0
0	1	1	1	1	1	1	1
1	0	0	0	1	1	1	1
1	0	1	0	1	1	1	1
1	1	0	0	1	1	1	1
1	1	1	1	1	1	1	1

Now columns 5 and 8 in the truth table 3.12 are identical. Therefore the two sides of the equation will yield the same results.

3.8   Simplification of Logic Functions

Now consider the simplification of some basic logic functions, using both the algebraic method and the switching diagram method.

Example 3.6
Verify

$$
A + AB = A
$$ (3.37)

Solution
Algebraically,

$$
A + AB = A(1 + B) = A \qquad \text{(by Equation 3.16 and 3.12)}
$$

or

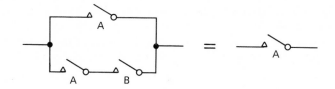

Example 3.7
Verify

$$A(A + B) = A \tag{3.38}$$

Solution
Algebraically,

$$A(A + B) = AA + AB = A + AB = A \qquad \text{(by Equation 3.13 and 3.37)}$$

or

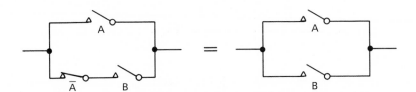

Example 3.8
Verify

$$A + \overline{A}B = A + B \tag{3.39}$$

Solution
Consider the diagrammatic solution first. Diagrammatically,

Sometimes there is more than one algebraic solution. Some of these will now be examined.

(i) $A + \overline{A}B = A(1 + B) + \overline{A}B$ by Equations 3.12 and 3.16)

$\qquad\qquad\quad = A + AB + \overline{A}B$

$\qquad\qquad\quad = A + B(A + \overline{A})$

$\qquad\qquad\quad = A + B$ (by Equations 3.18 and 3.12)

(ii) $A + \overline{A}B = \overline{A + \overline{A + \overline{B}}}$ (by De Morgan's theorem)

$\qquad\qquad\quad = \overline{\overline{A}(A + \overline{B})}$ (by De Morgan's theorem)

$\qquad\qquad\quad = \overline{\overline{A}A + \overline{A}\overline{B}}$ (by multiplication)

$\qquad\qquad\quad = \overline{0 + \overline{A}\overline{B}}$

$\qquad\qquad\quad = \overline{\overline{A}\overline{B}} = A + B$ (by De Morgan's theorem)

(iii) Multiplying the right hand side of the Equation 3.39 by Equation 3.18, i.e. $(A + \overline{A}) = 1$, gives

$$
\begin{aligned}
A + B &= (A + B)(A + \overline{A}) \\
&= AA + A\overline{A} + AB + \overline{A}B \\
&= A + 0 + AB + \overline{A}B &&\text{(by Equation 3.14)} \\
&= A(1 + B) + \overline{A}B &&\text{(combining first and third terms)} \\
&= A + \overline{A}B &&\text{(by Equation 3.16)}
\end{aligned}
$$

Example 3.9
Verify

$$B(A + \overline{B}) = AB \tag{3.40}$$

Solution
Algebraically

$$
\begin{aligned}
B(A + \overline{B}) &= AB + B\overline{B} \\
&= AB &&\text{(since } B\overline{B} = 0, \text{ by Equation 3.14)}
\end{aligned}
$$

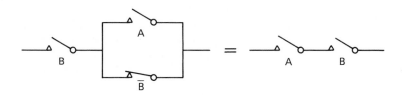

Example 3.10
Verify

$$AB + AC = A(B + C) \tag{3.41}$$

Solution
Diagrammatically

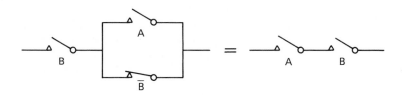

There is no need for an algebraic solution this time. It is a reverse of the distributive law given by Equation 3.33.

Note that during multiplication and most other algebraic operations, the variables are taken in alphabetical order. This helps the recognition of similar terms, rearranging them, combining terms with common factors, and so on.

Example 3.11
Verify

$$(A + B)(B + C)(C + \overline{A}) = (A + B)(C + \overline{A}) \tag{3.42}$$

Solution
Diagrammatically,

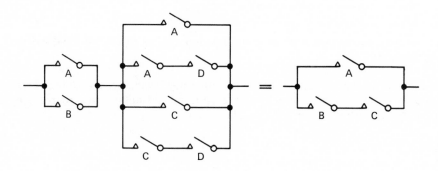

The two switches marked B are in series, and can therefore be replaced by a single switch B. Similarly the two switches C are replaced by a single switch C.

   Equation 3.42 is verified algebraically by multiplying the second term by the first term and then multiplying the result by the third term.

$$
\begin{aligned}
(A + B)(B + C)(C + \bar{A}) &= (AB + AC + BB + BC)(C + \bar{A}) \\
&= ABC + ACC + BBC + BCC + A\bar{A}B + A\bar{A}C + \\
&\quad \bar{A}BB + \bar{A}BC \\
&= ABC + AC + BC + 0 + 0 + \bar{A}B + \bar{A}BC \\
&= AC(B + 1) + BC + \bar{A}B(1 + C) \\
&= AC + BC + \bar{A}B \\
&= AC + BC + \bar{A}A + \bar{A}B \qquad \text{(by adding zero, i.e.} \\
&\qquad\qquad\qquad\qquad\qquad\qquad 0 = \bar{A}A) \\
&= C(A + B) + \bar{A}(A + B) \\
&= (A + B)(C + \bar{A})
\end{aligned}
$$

The algebraic solution shows that the term $(B + C)$ is redundant since B is covered by the first term and C is covered by the third term. The electrical circuit clearly emphasises it.

Example 3.12
Shaw that terms made up of several variables are redundant if one or more of these variables appear as separate terms in that expression. For instance, show that

$$(A + B)(A + AD + C + CD) = A + BC \tag{3.43}$$

Solution
Diagrammatically,

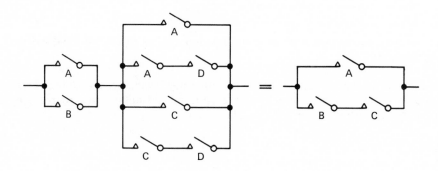

Algebraically,

$$(A + 3)(A + AD + C + CD) = (A + B)(A(1 + D) + C(1 + D))$$
$$= (A + B)(A + C) \quad \text{(since } 1 + D = 1)$$
$$= AA + AC + AB + BC$$
$$= A + AC + AB + BC$$
$$= A(1 + B + C) + BC$$
$$= A + BC$$

3.9 Summary of useful identities

A list of useful identities is given below.

$$A\overline{A} = 0$$
$$A + 1 = 1$$
$$\overline{A} + 1 = 1$$
$$\overline{B} + 1 = 1$$
$$\overline{A}(B + 1) = \overline{A}$$
$$\overline{A}(B + \overline{B}) = \overline{A}$$

These are used as terms or factors to add to, or to multiply the existing expressions by, in order to rearrange the function to obtain the required result. Some of these expressions were used in solving the identities is section 3.8; others will be used later.

3.10 Functions of two variables

Now consider a derivation of all the possible functions of two variables. As was shown previously, for two variables A and B there are four possible combinations of switching states which are represented by 1's and 0's. These are shown in Table 3.13.

Table 3.13
Switching
states for
two
variables

A	B
0	0
0	1
1	0
1	1

For n variables, each having two switching states, there are 2^{2^n} possible switching combinations. Therefore the variables A and B yield 16 different functions, as shown in Table 3.14.

Table 3.14 in fact represents a combination of 16 truth tables providing 16 different functions. All these functions may be derived from the basic set of logical expressions NOT, AND and OR.

It will be shown in the following chapters that all logical operations may be performed using the three functions NOT, AND and OR, plus the time variable (delay). It must be made clear, however, that logic algebra excludes time delays merely representing logical relationships.

Some of the functions in Table 3.14 may require verification. These are:

Table 3.14   Functions of two variables

0	0	1	1	A
0	1	0	1	B
0	0	0	0	$F_0 = 0$ (never)
0	0	0	1	$F_1 = AB$ (AND)
0	0	1	0	$F_2 = A\bar{B}$ (A but NOT B)
0	0	1	1	$F_3 = A$
0	1	0	0	$F_4 = \bar{A}B$ (B but NOT A)
0	1	0	1	$F_5 = B$
0	1	1	0	$F_6 = A\bar{B} + \bar{A}B$ (non-equivalence $A \neq B$, exclusive OR, $A \oplus B$)
0	1	1	1	$F_7 = A + B$ (OR)
1	0	0	0	$F_8 = \bar{A}\bar{B}$ (NOR, $\overline{A + B}$)
1	0	0	1	$F_9 = \bar{A}\bar{B} + AB$ (equivalence $A \equiv B$)
1	0	1	0	$F_{10}\ \bar{B}$ (NOT B)
1	0	1	1	$F_{11} = A + \bar{B}$ (B implies A)
1	1	0	0	$F_{12} = \bar{A}$ (NOT A)
1	1	0	1	$F_{13} = \bar{A} + B$ (A implies B)
1	1	1	0	$F_{14} = \bar{A} + \bar{B}$ (NAND, $\overline{AB}$)
1	1	1	1	$F_{15} = 1$ (always)

$$F_3 = A\bar{B} + AB = A(\bar{B} + B) = A$$
$$F_5 = \bar{A}B + AB = B(\bar{A} + A) = B$$
$$F_7 = \bar{A}B + A\bar{B} + AB = (\bar{A} + A)B + A\bar{B} = B + A\bar{B} = A + B$$
$$F_{10} = \bar{A}\bar{B} + A\bar{B} = \bar{B}(\bar{A} + A) = \bar{B}$$
$$F_{11} = \bar{A}\bar{B} + A\bar{B} + AB = \bar{B}(\bar{A} + A) + AB = \bar{B} + AB = A + \bar{B}$$
$$F_{12} = \bar{A}\bar{B} + \bar{A}B = \bar{A}(\bar{B} + B) = \bar{A}$$
$$F_{13} = \bar{A}\bar{B} + \bar{A}B + AB = \bar{A}(\bar{B} + B) + AB = \bar{A} + AB = \bar{A} + B$$
$$F_{14} = \bar{A}\bar{B} + \bar{A}B + A\bar{B} = \bar{A}(\bar{B} + B) + A\bar{B} = \bar{A} + A\bar{B} = \bar{A} + \bar{B}$$

3.11   Derivation of a function

Consider the three basic steps taken in logical design.

(i)   The problem is defined in terms of inputs and outputs.
(ii)   A truth table is derived to cover all possible combinations of the variables involved.
(iii)   All variables, i.e. A, B, C, . . . terms, are extracted from the truth table which give the function F = 1.

Example 3.13
Consider the classic example of switching a light *on* and *off* using one of the two available switches.

Solution
Let A and B represent the two switches. Then

$$A = 0 \text{ represents switch A } off$$
$$B = 0 \text{ represents switch B } off$$
$$A = 1 \text{ represents switch B } on$$
$$\text{and } B = 1 \text{ represents switch B } on$$

Now, let F represent the light condition. Then

$$F = 0 \text{ represents light } off, \text{ and}$$
$$F = 1 \text{ represents light } on.$$

The switching function, F, is characterised by the truth table 3.15.

Table 3.15   Truth table for Example 3.13

A	B	F
0	0	0
0	1	1
1	0	1
1	1	0

Extracting from the truth table 3.15 those A and B terms which give $F = 1$, we get

$$F = \overline{A}B + A\overline{B} \tag{3.44}$$

which is the exclusive-OR function.

Equation 3.44 is in the form of a **sum-of-products**, i.e. a logical sum of logical products.

Extracting from the truth table those A and B terms which give $F = 0$, an inverted function is obtained.

$$\overline{F} = \overline{A}\overline{B} + AB \tag{3.45}$$

Inverting Equation 3.45 gives

$$F = \overline{\overline{A}\overline{B} + AB} \tag{3.46}$$

and applying De Morgan's law gives

or $F = (A + B)(\overline{A} + \overline{B})$ $\tag{3.47}$

which is in the form of a **product of sums**.

This shows that any function may be expressed as a sum-of-products or as a product-of-sums. We will consider this in more detail

3.12   Minterm and maxterm forms

The sum-of-products and product-of-sums forms were mentioned in Example 3.13. A general procedure will now be examined for obtaining a switching function.

Recalling, from Equation 3.3, that the AND operation has the symbolic form of product, if each line of the truth table is expressed in its symbolic form a column of **minterms** is obtained for the two variables. This is illustrated in Table 3.16.

Table 3.16   Minterms for two variables

A	B	Minterms
0	0	$\overline{A}\overline{B}$
0	1	$\overline{A}B$
1	0	$A\overline{B}$
1	1	AB

Similarly, the minterms for *three* variables are shown in Table 3.17.

This procedure may be extended to any number of variables.

A **minterm** is defined as a **switching product**, i.e. variables ANDed together, and the sum-of-products is called the **minterm form** of the function.

Table 3.17 Minterms for three variables

A	B	C	Minterms
0	0	0	$\bar{A}\bar{B}\bar{C}$
0	0	1	$\bar{A}\bar{B}C$
0	1	0	$\bar{A}B\bar{C}$
0	1	1	$\bar{A}BC$
1	0	0	$A\bar{B}\bar{C}$
1	0	1	$A\bar{B}C$
1	1	0	$AB\bar{C}$
1	1	1	ABC

Table 3.18 Maxterms for two variables

A	B	Maxterms
0	0	$\bar{A} + \bar{B}$
0	1	$\bar{A} + B$
1	0	$A + \bar{B}$
1	1	$A + B$

Table 3.19 Maxterms for three variables

A	B	C	Maxterms
0	0	0	$\bar{A} + \bar{B} + \bar{C}$
0	0	1	$\bar{A} + \bar{B} + C$
0	1	0	$\bar{A} + B + \bar{C}$
0	1	1	$\bar{A} + B + C$
1	0	0	$A + \bar{B} + \bar{C}$
1	0	1	$A + \bar{B} + C$
1	1	0	$A + B + \bar{C}$
1	1	1	$A + B + C$

Now consider Equation 3.6, A & B. The expression A + B is called a **maxterm**. Again a column of maxterms for the two variables can be formed as shown in Table 3.18.

For three variables, the maxterms are shown in Table 3.19.

Thus a **maxterm** is defined as a **logical sum**, or as variables ORed together. The product-of-sums is called the **maxterm form** of the function.

It is thus possible to represent any function in the **standard sum** form, which is the logical sum of all the minterms describing the particular function, or as the **standard product** form which is the logical product of all the maxterms describing the particular function.

3.13 Canonical form

The **canonical** form refers to a method of representing all the variables, inverted or non-inverted, once only. This form of a function is used for the comparison of functions term by term, and for Karnaugh mapping, which is described in Chapter 5.

If there are any missing variables in an expression, then the algebraic method of expansion for minterms is used. This involves multiplying logically each term by the absent variables expressed in the form $(A + \bar{A})$, $(B + \bar{B})$, etc.

In general, for n binary variables there are 2^n minterms and 2^n maxterms. For instance, if $n = 3$, there are 8 minterms and 8 maxterms. Note the following points.

(i) The logical sum of all the minterms is 1, i.e.
$$\bar{A}\bar{B}\bar{C} + \bar{A}\bar{B}C + \bar{A}B\bar{C} + \bar{A}BC + A\bar{B}\bar{C} + A\bar{B}C + AB\bar{C} + ABC = 1 \qquad (3.48)$$

(ii) The logical product of all maxterms is 0, i.e.
$$(A + B + C)(A + B + \bar{C})(A + \bar{B} + C)(A + \bar{B} + \bar{C})(\bar{A} + B + C)$$
$$(\bar{A} + B + \bar{C})(\bar{A} + \bar{B} + C)(\bar{A} + \bar{B} + \bar{C}) = 0 \qquad (3.49)$$

(iii) The complement of any minterm is a maxterm.

(iv) The complement of any maxterm is a minterm.

Example 3.14

Derive the canonical form of the function $F = A\bar{C} + \bar{B}\bar{C}$.

Solution

Since F is a three variable function it has to be expanded by logically multiplying the first term by $(B + \bar{B})$ and the second term by $(A + \bar{A})$. Thus

$$F = A\bar{C}(B + \bar{B}) + \bar{B}\bar{C}(A + \bar{A})$$
$$= AB\bar{C} + A\bar{B}\bar{C} + A\bar{B}\bar{C} + \bar{A}\bar{B}\bar{C} \qquad (3.50)$$

Note that the alphabetical order of the variables in each term is always maintained to facilitate simplification. Also all repeated terms are eliminated. Therefore the canonical form of the function $F = A\bar{C} = \bar{B}\bar{C}$ is

$$F = AB\bar{C} + A\bar{B}\bar{C} + \bar{A}\bar{B}\bar{C} \qquad (3.51)$$

The solution to Example 3.14 is in the form of the sum-of-products. Suppose the truth table is written for the function as shown in Table 3.20. From the truth table, writing the minterms for each 0 of the function gives

$$\bar{F} = \bar{A}\bar{B}C + \bar{A}B\bar{C} + \bar{A}BC + A\bar{B}C + ABC \qquad (3.52)$$

Table 3.20 Truth table for $F = AB\bar{C} + A\bar{B}\bar{C} + \bar{A}\bar{B}\bar{C}$

A	B	C	F
0	0	0	1
0	0	1	0
0	1	0	0
0	1	1	0
1	0	0	1
1	0	1	0
1	1	0	1
1	1	1	0

To obtain F, the complement of the function is taken, i.e. the complement of both sides of Equation 3.52. Therefore

$$F = \overline{\bar{A}\bar{B}C + \bar{A}B\bar{C} + \bar{A}BC + A\bar{B}C + ABC} \qquad (3.53)$$

Applying De Morgan's theorem to the right hand side of Equation 3.53 gives

$$F = (\overline{\bar{A}\bar{B}C})\,(\overline{\bar{A}B\bar{C}})\,(\overline{\bar{A}BC})\,(\overline{A\bar{B}C})\,(\overline{ABC})$$

or

$$F = (A + B + \bar{C})(A + \bar{B} + C)(A + \bar{B} + \bar{C})(\bar{A} + B + \bar{C})(\bar{A} + \bar{B} + \bar{C}) \qquad (3.54)$$

Equation 3.54 is in the form of a product of maxterms, otherwise known as the **canonical product of sums**.

3.14 Problems

3.1 Write the truth table for a NOT function.

3.2 Write the truth table for a two-variable AND function.

3.3 Repeat Problem 3.2 for a three-variable function.

3.4 Write the truth table for a two-variable OR function.

3.5 Repeat Problem 3.4 for a three-variable function.

3.6 Write a truth table for an exclusive-OR function.

3.7 Draw a switching diagram to represent the exclusive-OR function.

3.8 Draw a switching diagram to represent the following function, $F = AC + \overline{A}B + BC$.

3.9 Write the truth table for the following logical expression, $F = AB + (A + C)(B + C)$.

3.10 Write the truth table for the expression $F = (A + B)(BC)(A + C)$.

3.11 Simplify the following functions:

(a) $F = A(A + B)$ (b) $F = A(\overline{A} + B)$ (c) $F = AB + A\overline{B}$
(d) $F = AB + AC + B\overline{C}$ (e) $F = A + AB + \overline{A}B$
(f) $F = AB + A\overline{B} + \overline{A}B + \overline{A}\overline{B}$ (g) $F = (A + B)(A + \overline{B})$
(h) $F = (A + B)(B + C)(\overline{A} + C)$

3.12 Write the truth tables for the functions shown in Problem 3.11, and for their simplified versions.

3.13 Write the minterms for a switching function of the three variables A, B and C.

3.14 Derive the canonical form for the following functions:

(a) $F = A + A\overline{B}$ (b) $F = AB + BC$ (c) $F = A + AB + AC$
(d) $F = AC + \overline{A}B + \overline{B}\overline{C}$

3.15 Write the maxterms for the switching functions in Problem 3.14.

3.16 Write the truth table for the expression $F = \overline{(\overline{A} + \overline{B}) + (\overline{A} + \overline{C}) + (\overline{B} + \overline{C})}$.

3.17 Write the truth table for the expression $F = \overline{\overline{AB} + \overline{AC} + \overline{BC}}$.

3.18 In the following truth table, the function F is given in terms of the three variables A, B and C. Express F as a canonical sum-of-products assuming that all variables and their complements are available.

A	B	C	F
0	0	0	0
0	0	1	1
0	1	0	1
0	1	1	0
1	0	0	1
1	0	1	0
1	1	0	0
1	1	1	1

3.19 Express the function F in Problem 3.18 as a canonical product-of-sums.

4

COMBINATIONAL LOGIC

4.1 Introduction

This chapter is concerned with the realisation in circuit form of the logical functions investigated in Chapter 3. Specific circuits knows as **gates** are used to do this. Each gate is represented by a symbol corresponding to the appropriate logic function realised by that specific gate. Gates can be constructed from discrete components such as transistors, resistors and diodes, but are now always implemented in integrated circuit form. The structure of these gates is investigated in Chapter 6, and this chapter will treat gates (integrated circuit only) as functional units. There are various ways of producing a particular logic function by different arrangements or combination of gates.

Combination logic implies the use of logic gates to produce logical functions that are non-time-dependent. Later chapters will deal with time-dependent circuits which fall into the category of sequential logic. The chapter will start by considering the range of gates available, and then will show how these gates can be interconnected to form more complex functions.

4.2 Basic logic elements and symbols

The logical functions described in Chapter 3 will now be implemented using gates. The gates to be considered have a minimum of two inputs and one output, with exception of the NOT gate, or **inverter**, which has only one input and one output. The number of inputs and outputs depends on the number of variables used and the complexity of the function. There are, infact, no theoretical limitations to the number of inputs that a gate can have. From a practical point of view, however, the limit imposed is that due to the number of pins available in the integrated circuit package. Some examples of the commercially available 4000 and 7400 series and the new high speed low power CMOS family in 14-pin packages are as follows: six one-input gates, four two-input gates, three three-input gates, two four-input gates and one eight-input gate.

The 7400 TTL family of devices are manufactured by Fairchild, Hitachi, Motorola, Philips/Signetics, National Semiconductor, and Texas Instruments. The 4000 CMOS family of devices are manufactured by Motorola, Philips/Signetics, National Semiconductors and RCA. The characteristics and operating conditions of logic families are discussed in Chapter 6 and some examples of the integrated circuit packages are given in Chapter 11.

Many sets of logic symbols have been developed over the past twenty years, but there are only two sets of symbols in current use which are worth considering. These are the American Military Standard and the British Standard 3939. Since most manufacturers' data, books, technical, journals and industry itself follow the American Military Standard, it will also be used in this book. This standard and equivalent British Standard symbols of the most commonly used gates are shown in Fig. 4.1.

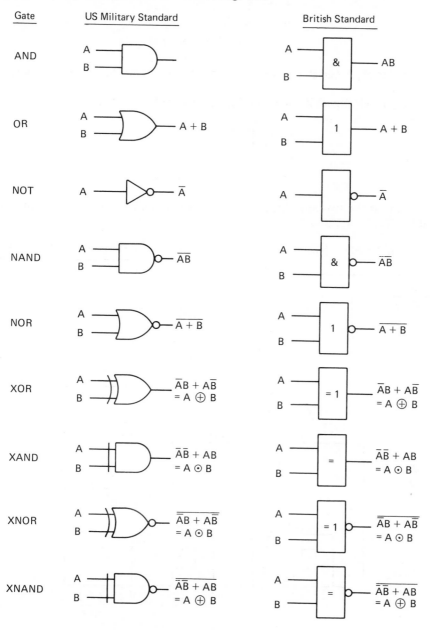

Fig 4.1 Commonly used logic symbols

4.3 The AND, OR and NOT gates

AND gates

A logic circuit that performs the AND operation is called an **AND gate**. The symbol for a two-input AND gate is shown in Fig. 4.2, where A and B are the input variables and F is the output. The function is given by F = AB

A —
B —
F = AB

Fig 4.2   Two-input AND gate

The input variables, A and B, can take the form of a logic 0 or a logic 1. F is a 1 only if both A and B are a 1.

A three-input AND gate is shown in Fig. 4.3, where F = ABC.

A —
B —
C —
F = ABC

Fig 4.3   Three-input AND gate

The operation of the two and three-input AND gates can be summarised in the form of truth tables (Tables 4.1 and 4.2).

Table 4.1   Truth table for a 2-input AND gate

A	B	F
0	0	0
0	1	0
1	0	0
1	1	1

Table 4.2   Truth table for a three-input AND gate

A	B	C	F
0	0	0	0
0	0	1	0
0	1	0	0
0	1	1	0
1	0	0	0
1	0	1	0
1	1	0	0
1	1	1	1

V_{CC} = +5 V

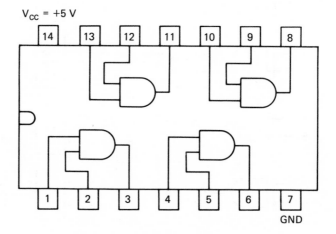

Fig 4.4   7408 quad two-input AND gate

The 0 and 1 logic levels are physically represented by voltages, as described in Chapter 1. Input and output logic levels can therefore be monitored by using a voltmeter. A typical integrated circuit containing four AND gates is the 7408, illustrated in Fig. 4.4.

Figure 4.4 illustrates the way in which the four AND gates are connected to the pins on the 14-pin dual-in-line (DIL) integrated circuit (IC). Note that the IC gates are active devices, requiring a power supply – in this case 5 V and 0 V as shown.

OR gates

A logic circuit that performs the logical OR operation is an OR gate. The symbols for a two input and three-input OR gate are shown in Figs 4.5(a) and (b) respectively.

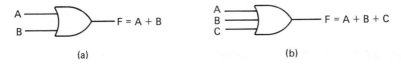

(a)   (b)

Fig 4.5   (a) Two-input OR gate; (b) three-input OR gate

The truth tables for gates of Fig. 4.5 (Tables 4.3 and 4.4) indicate that the output of an OR gate is a logic 1 if any one or more inputs are logic 1.

Table 4.3   Truth table for $F = A + B$

A	B	F
0	0	0
0	1	1
1	0	1
1	1	1

Table 4.4   Truth table for $F = A + B + C$

A	B	C	F
0	0	0	0
0	0	1	1
0	1	0	1
0	1	1	1
1	0	0	1
1	0	1	1
1	1	0	1
1	1	1	1

Multiple-input AND and OR

Both and AND and the OR operation can be used with more than three inputs. Figure 4.6 shows examples of the symbols used for multiple input gates.

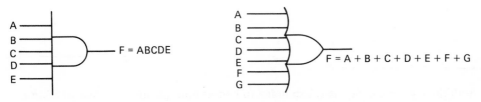

Fig 4.6   Symbols for multiple input gates

NOT gates

The logical NOT operation (complementation or inversion) is represented symbolically by means of a NOT gate, as illustrated in Fig. 4.7. The output of the NOT gate always assumes the opposite state to its input state, as shown in Table 4.5. The small circle at the output is a standard way of representing an inversion. It will be used in conjunction with other gates described in the following paragraphs.

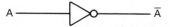

Fig 4.7   The symbol for a NOT gate

Table 4.5   Truth table for a NOT gate

A	$\bar{A}$
0	1
1	0

4.4 The NAND and NOR gates

Two frequently used gates are the NAND and NOR gates. Their popularity stems from the fact that the circuit elements used to implement these logic functions automatically invert the logical result, which makes it possible to derive any of the logic gates from either all NAND or all NOR gates.

The NAND gate symbolically realises the NOT–AND operation, as shown in Fig. 4.8.

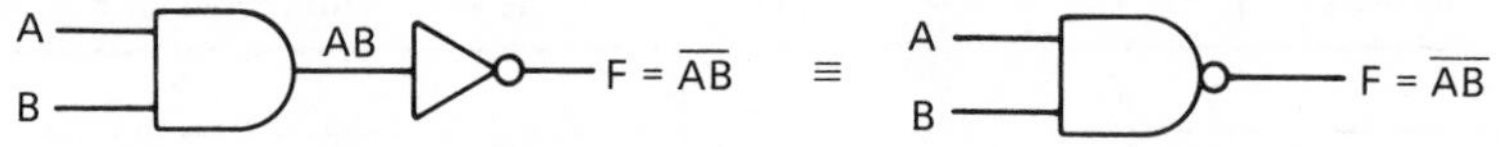

Fig 4.8   NAND gate operation and NAND gate symbol

Similarly the NOR gate realises the NOT–OR operation as illustrated in Fig. 4.9.

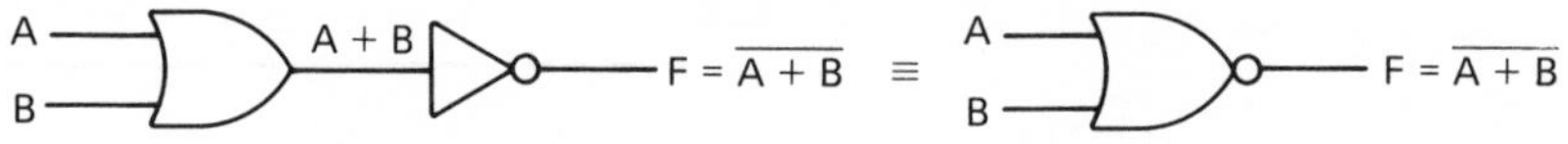

Fig 4.9   NOT-OR operation and NOR gate symbol

The truth tables for the NAND gate and the NOR gate are given in Tables 4.6 and 4.7 respectively.

Table 4.6   Truth table for NAND gate

A	B	F
0	0	1
0	1	1
1	0	1
1	1	0

Table 4.7   Truth table for NOR gate

A	B	F
0	0	1
0	1	0
1	0	0
1	1	0

Both these gates can be easily expanded to handle multiple inputs. Their operations are illustrated symbolically in Fig. 4.10 and 4.11.

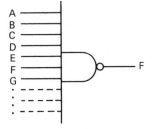

Fig 4.10 Multi-input NAND gate

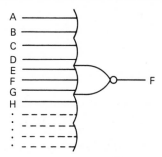

Fig 4.11 Multi-input NOR gate

4.5 Forcing function

An extremely useful approach to using simple logic gates is to consider the **forcing function** of each type. A forcing function will either be a logic 1 or a logic 0. It is the single input to a gate that will predetermine the output regardless of any other inputs. An example will make this clearer.

The truth table and symbol for an AND gate are shown in Fig. 4.12. With A = 0, the input to B is irrelevant. This is shown in Fig. 4.12(b) as B = X, which is the "Don't Care" condition, i.e. B = 0 or B = 1, as long as A = 0, the output A.B will always be 0. In other words a 0 on either A or B (or both) will force a 0 out so the forcing function is 0.

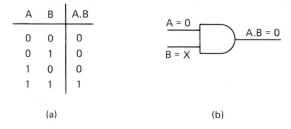

A	B	A.B
0	0	0
0	1	0
1	0	0
1	1	1

(a)   (b)

Fig 4.12 Forcing function of an AND gate

If a NAND gate is considered, as in Fig. 4.13, the forcing function is still 0 but the output is now 1.

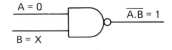

Fig 4.13 Forcing function of a NAND gate

If an OR gate is considered a 1 on either input (or both) will force a 1 output, as in Fig. 4.14. The forcing function for an OR gate is therefore a logic 1.

Similarly, for a NOR gate (Fig. 4.15), the forcing function is 1 with a 0 output.

The advantages of using the forcing function concept as are follows.

(i)   There is no need to memorise truth tables for the four basic gates shown,
       (a)   With AND, forcing function = 0, output = 0

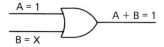

Fig 4.14   Forcing function of an OR gate

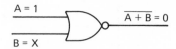

Fig 4.15   Forcing function of a NOR gate

(b)   With NAND, forcing function = 0, output = 1
(c)   With OR, forcing function = 1, output = 1
(d)   With NOR, forcing function = 1, output = 0
For example, if it is remembered that 1 is the forcing function for a NOR gate, then whenever a 1 is present at the input a 0 is fed out. If there are no 1's at the input, a 1 will be fed out.

(ii)   When testing gate circuits using a logic probe (a device which indicates logic levels) on a gate with a number of inputs, detection of a forcing function on any input will immediately indicate what the output level should be, *without having to test the remaining inputs*.
If a disparity is found then a fault is present and can be rectified.

4.6   Implementation of the NOT gate

Before embarking on the implementation of algebraic espressions, the importance of NAND and NOR gates must be emphasised.

Both these gates can be very easily converted to a NOT gate. This opens a completely new field of applications, where the entire network can be implemented with only NAND gates or with only NOR gates. For example, if a two-input NAND gate has the same variable A applied to to both inputs, then the output F becomes $\overline{A}$, as illustrated in Fig. 4.16. This is equivalent to a NOT gate (an inverter).

$$F = \overline{AA} = \overline{A} \equiv \overline{A}$$

Fig 4.16   NOT gate using a two-input NAND gate

Similarly, a two-input NOR gate can be converted to a NOT gate by the application of the same variable to both inputs. This time a variable B is used, as shown in Fig. 4.17. Again the result is an inverter.

$$F = \overline{B + B} = \overline{B} \equiv \overline{B}$$

Fig 4.17   NOT gate using a two-input NOR gate

The above two proofs can be extended to the multiple-input NAND and NOR gates. No matter how many inputs are available, if all the inputs are strapped together the result is a single input gate producing the NOT function. This is symbolically represented in Fig. 4.18 and 4.19.

A
A
A
A
A
F = $\overline{AA \cdots A}$
= $\overline{A}$
≡
A
$\overline{A}$

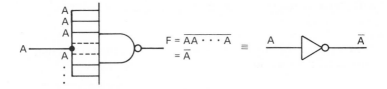

Fig 4.18 Multiple input NAND gate conversion to a NOT gate

B
B
B
B
B
F = $\overline{B + B + \cdots B}$
= $\overline{B}$
≡
B
$\overline{B}$

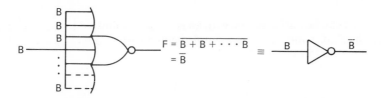

Fig 4.19 Multiple input NOR gate conversion to a NOT gate

4.7 Implementation of algebraic expressions

Any algebraic expression can be implemented using the AND, OR and NOT gates previously described.

Example 4.1
Implement both terms of De Morgan's first theorem, which states that the complement of the logical sum equals the logical product of the complements.

Solution
De Morgan's first theorem can be represented by

$$\overline{A + B} = \overline{A}\,\overline{B}$$

The left hand side can be implemented by an OR gate followed by an inverter, as shown in Fig. 4.20. The right hand side can be implemented by two NOT gates and an AND gate, as shown in Fig. 4.21.

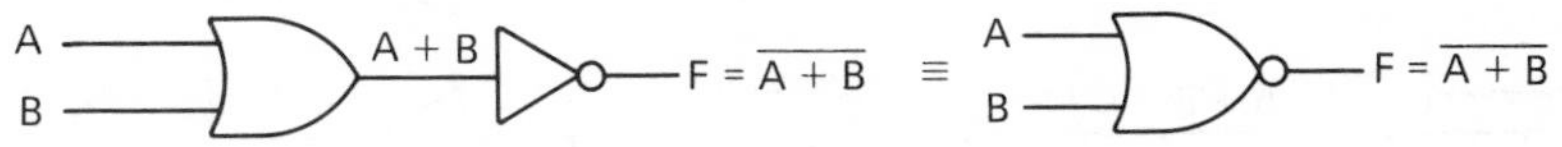

A
B
A + B
F = $\overline{A + B}$
≡
A
B
F = $\overline{A + B}$

Fig 4.20 Implementation of $\overline{A+B}$

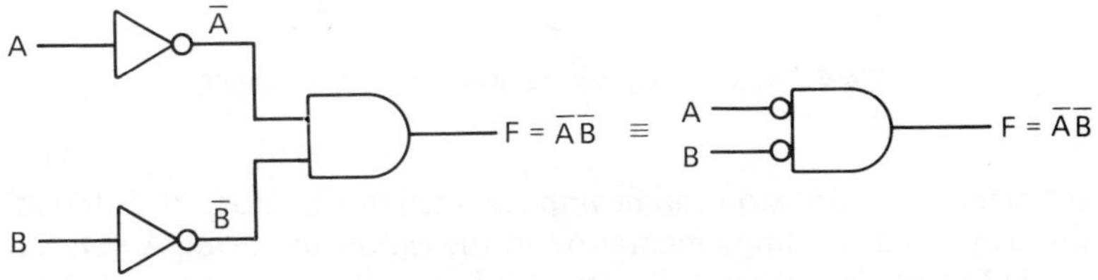

A
$\overline{A}$
B
$\overline{B}$
F = $\overline{A}\,\overline{B}$
≡
A
B
F = $\overline{A}\,\overline{B}$

Fig 4.21 Implementation of $\overline{A}\,\overline{B}$

Thus a NOR gate is equivalent to an AND gate with inverted inputs.

Example 4.2
Implement both terms of De Morgan's second theorem, which states that the complement of a logical product equals the logical sum of the complements.

Solution
De Morgan's second theorem can be represented by

$$\overline{AB} = \overline{A} + \overline{B}$$

The left hand side can be implemented with an AND gate followed by an inverter, as shown in Fig. 4.22. The right hand side can be implemented with two NOT gates followed by an OR gate, as illustrated in Fig. 4.23.

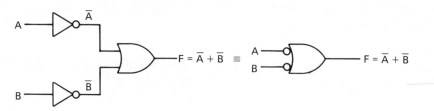

Fig 4.22   Implementation of $\overline{AB}$

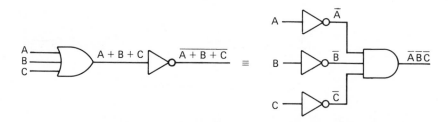

Fig 4.23   Implementation of $\overline{A} + \overline{B}$

Thus a NAND gate is equivalent to an OR gate with inverted inputs.
   The two De Morgan's theorems can be extended to more variables. The implementations for three variables are shown in Fig. 4.24 and 4.25.

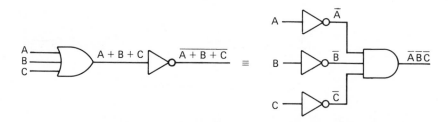

Fig 4.24   De Morgan's first theorem for three inputs

In fact any algebraic expression can be implemented in a number of different ways. The most obvious way is a direct implementation of the expression using AND, OR and NOT gates. Using De Morgan's theorems, however, it is possible to transform the expression so that any desired or available set of gates will produce the required result.

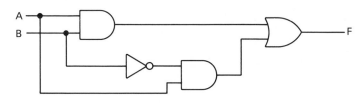

Fig 4.25   De Morgan's second theorem for three variables

Example 4.3
Given the circuit shown in Fig. 4.26, find F.

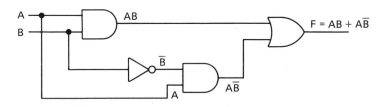

Fig 4.26   Circuit for Example 4.3

Solution
Starting from the left hand side, write the output function of each gate, gradually working to obtain the final expression for F. This is shown in Fig. 4.27.

Fig 4.27   Solution to Example 4.3

The output $F = AB + A\overline{B}$, however, simplifies to $F = A(B + \overline{B}) = A$. This example shows that, to avoid unnecessary gates, simplification is necessary before any function is implemented.

Since $F = A$, the variable B is not required and, in fact, no other gate is needed since direct connection from A to F satisfies the function.

4.8   Gate equivalents

A logic function, whether in the product-of-sums or sum-of-products form, is expressed in terms of AND, OR and NOT. However, standardised logic integrated circuits are commonly in NAND or NOR form. Therefore it is frequently necessary to convert from

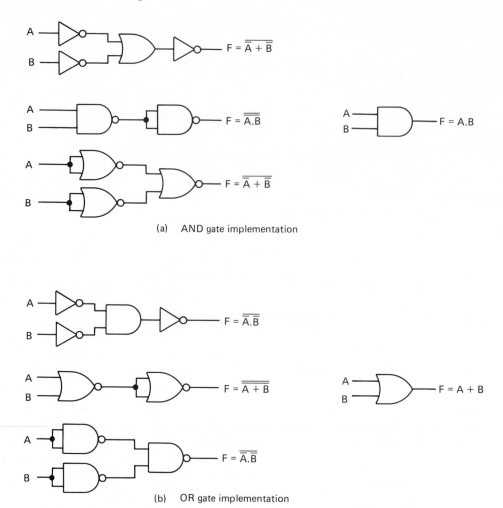

(a)   AND gate implementation

(b)   OR gate implementation

(c)   NAND from NOR

(d)   NOR from NAND

Fig 4.28   (a) AND gate implementation; (b) OR gate implementation; (c) NAND from NOR; (d) NOR from NAND

AND/OR to NAND or NOR expressions. To illustrate this, the following gates will be implemented:

(a) AND gate
  (i)   using OR and NOT gates
  (ii) using NAND gates
  (iii) using NOR gates
(b) OR gate
  (i)   using AND and NOT gates
  (ii) using NOR gates
  (iii) using NAND gates
(c) NAND gate using NOR gates
(d) NOR gate using NAND gates

These are illustrated in Fig. 4.28.

4.9 Exclusive OR logic implementation

The Exclusive OR (XOR) operation, defined in Chapter 3 as $F = \overline{A}B + A\overline{B}$ or $F = A \oplus B$, is represented symbolically in Fig. 4.29.

A
B
$F = A \oplus B$

Fig 4.29 Exclusive-OR gate

The operation of the gate is very similar to the OR gate except that, as seen from the truth table 4.8, the output is a 0 when both inputs are a 1. The output from an OR gate is high with a 1 on both inputs, and can be called an inclusive-OR. The exclusive-OR excludes this condition.

Table 4.8 Truth table for XOR gate

A	B	F
0	0	0
0	1	1
1	0	1
1	1	0

As can be seen from table 4.8, the output of a two-input exclusive-OR gate assumes a 1 state if one and only one input assumes a 1 state.

A three-input exclusive-OR gate represented by an expression $F = A \oplus B \oplus C$ will now be considered. This equation (using the associative law, described in Chapter 3) can be arranged in three different ways, indicating that the three-input XOR gate can be implemented with only two two-input XOR gates, as shown in Fig. 4.30, since $F = A \oplus B \oplus C = (A \oplus B) \oplus C = A \oplus (B \oplus C) = (A \oplus C) \oplus B$.

The truth table for the gate is shown in Table 4.9.

An expression for F from the truth table is

$$F = \overline{A}\overline{B}C + \overline{A}B\overline{C} + AB C + A\overline{B}\overline{C}$$

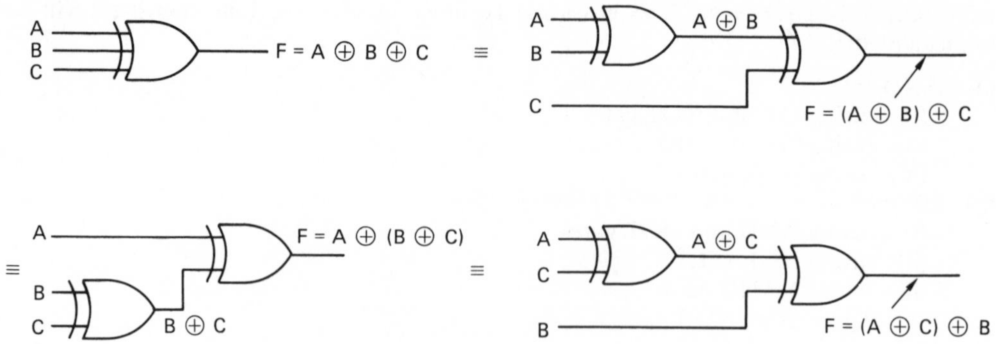

Fig 4.30   Three-input exclusive -OR gate

Table 4.9   Truth table for three-input XOR gates

A	B	C	F
0	0	0	0
0	0	1	1
0	1	0	1
0	1	1	0
1	0	0	1
1	0	1	0
1	1	0	0
1	1	1	1

To prove this identity, we take the first two-input implementation from Fig. 4.30 and transform algebraically to give:

$$F = (A \oplus B) \oplus C = (A\bar{B} + \bar{A}B)\bar{C} + \overline{(A\bar{B} + \bar{A}B)}C$$

$$= (A\bar{B} = \bar{A}B)\bar{C} = \overline{(A\bar{B} \cdot \bar{A}B)}\,C$$

$$= (A\bar{B} + \bar{A}B)\bar{C} + (\bar{A} + B)(A + \bar{B})C$$

$$= (A\bar{B} + \bar{A}B)\bar{C} + (\bar{A}\bar{B} + AB)C$$

$$= A\bar{B}\bar{C} + \bar{A}B\bar{C} + \bar{A}\bar{B}C + ABC$$

This has the same terms as the equation derived from the truth table. It is left to the reader to prove the identities for the remaining two implementations in Fig. 4.30.

In general, for any exclusive-OR gate the output is a 1 if any odd number of inputs are 1.

4.10   Exclusive-OR applications

One application of the exclusive-OR gate is in the ALU (Arithmetic and Logic Unit) of a computer. The design of logic circuits to perform arithmetic will be discussed in detail in Chapter 10, but the simple half-adder and half-subtractor circuit will be examined here to illustrate exclusive-OR applications.

Half Adder

To add two binary digits, a truth table is first written (Table 4.10) using the rules explained in Chapter 2. Then the logical expressions for sum and carry are derived.

Let the two binary digits be X and Y, then S is the sum and C is carry.

Table 4.10 Addition of two bits

X	Y	S	C
0	0	0	0
0	1	1	0
1	0	1	0
1	1	0	1

From Table 4.10, $S = \bar{X}Y + X\bar{Y}$ and $C = XY$. Therefore the circuit must produce a sum and carry facility at the same time. Both outputs can be obtained as shown in Fig. 4.31. The circuit in this form is referred to as a half adder because it does not permit a 'carry-in' input, i.e. it will only add *two* bits.

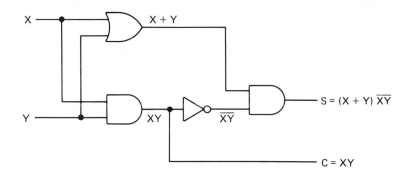

Fig 4.31 Half adder circuit

Another useful circuit to implement the half adder, which requires only NOR gates, is that of Fig. 4.32. The output for the carry is readily available, as seen from $\overline{\bar{X} + \bar{Y}} = XY = C$.

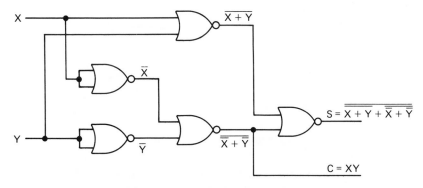

Fig 4.32 Half adder using NOR gates only

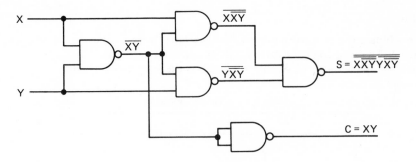

Fig 4.33   Half adder using NAND gates only

However, if NAND gates only are used, the circuit of Fig. 4.33 will produce the required functions.

Half subtractor

The arithmetic unit of a computer can be simplified, because subtraction can be performed using the same XOR circuit which is used for the half adder. A different output, however, is required to produce the borrow.

If the two bits to be substracted are X and Y, where D is the difference and B is borrow, then the truth Table 4.11 shows the results.

Table 4.11   Subtraction of two bits

X	Y	D	B
0	0	0	0
0	1	1	1
1	0	1	0
1	1	0	0

From Table 4.11, $D = \overline{X}Y + X\overline{Y}$ and $B = \overline{X}Y$. Note that the difference (D) is identical to the sum (S) in the half adder. To implement this, the circuit must produce the difference and borrow at the same time. A suitable circuit for this is shown in Fig. 4.34.

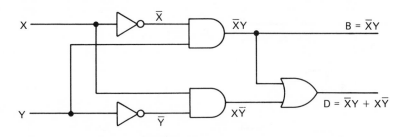

Fig 4.34   Half subtractor circuit

To implement a half subtractor with NOR gates only, the circuit of Fig. 4.35 is used.
To implement the half subtractor using NAND gates only, the circuit of Fig. 4.36 can be used.

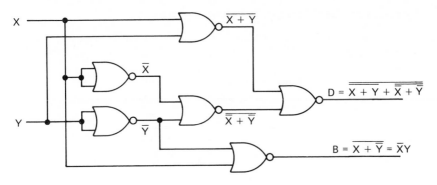

Fig 4.35 Half subtractor using NOR gates only

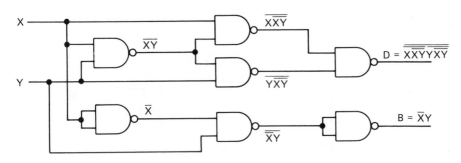

Fig 4.36 Half subtractor using seven NAND gates

In this circuit, however, two inversions are used to obtain the borrow output. It is possible to use a different version of the NAND only implementation with fewer NAND gates.

Since $D = \overline{X}Y + X\overline{Y}$, applying De Morgan's 2nd theorem gives $D = \overline{\overline{X}Y \, \overline{X\overline{Y}}}$, which can be implemented with five NAND gates. Only one additional invertor is needed to obtain the borrow output. The overall number of gates is reduced to six, as shown in Fig. 4.37.

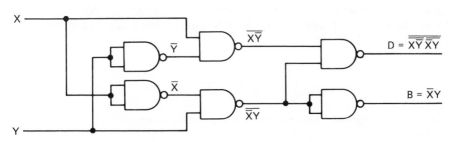

Fig 4.37 Half subtractor using six NAND gates

4.11   Exclusive NOR logic implementation

There are at least three names used for the XNOR logic gate: the exclusive-NOR, coincidence or equivalence gate. The XNOR logic gate is defined by the truth table 4.12.

Table 4.12   Truth table for XNOR gate

A	B	F
0	0	1
0	1	0
1	0	0
1	1	1

The XNOR is defined by the expression $F = AB + \overline{A}\overline{B}$, which can be implemented as shown in Fig. 4.38. The symbol for the XNOR gate is shown in Fig. 4.39.

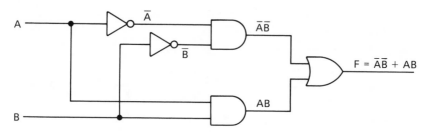

Fig 4.38   The coincidence gate (XNOR gate)

Fig 4.39   XNOR symbol

The profusion of terms used for describing the operation causes a certain amount of confusion for the newcomer to the subject. Unfortunately there is one more name associated with the same Boolean function, the exclusive-AND (XAND). The operation is defined by the same truth table as the XNOR gate (Table 4.12), but it has a separate symbol, as shown in Fig. 4.40.

$$F = A \odot B = \overline{A}\overline{B} + AB$$

A ⎯⎯
B ⎯⎯

Fig 4.40   The coincidence gate, exclusive-AND gate symbol

The XAND gate can be implemented using NAND gates as shown in Fig. 4.41, because $F = \overline{A}\overline{B} + AB = \overline{\overline{A}\overline{B} \ \overline{AB}}$. XAND is represented by the ringed product symbol $\odot$.

Combining all the forms discussed into a set of equivalent Boolean expressions, we get

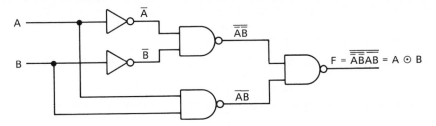

Fig 4.41 The coincidence gate (XAND) gate

$$F = \overline{A}\overline{B} + AB = \overline{\overline{A}B + A\overline{B}} = \overline{A \oplus B} = A \odot B$$

The coincidence function can be extended to three or more inputs since the ringed product operation obeys the association law described in chapter 3, so that

$$F = A \odot B \odot C = A \odot (B \odot C) = (A \odot B) \odot C$$
$$= \overline{A}\overline{B}C + \overline{A}B\overline{C} + A\overline{B}\overline{C} + ABC$$

From the truth table, it can be seen that F is also given by an inverse of the exclusive-OR function, i.e. $F_1 = \overline{A}B + A\overline{B}$, which is NOT exclusive OR, otherwise expressed as XOR or XNOR (exclusive-NOR). Applying De Morgan's theorem to F_1 gives

$$F_1 = \overline{\overline{A} + \overline{B}} + \overline{\overline{A} + B} = (A + B)(\overline{A} + B) = \overline{A}B + AB.$$

Therefore, implementing F_1 in the form $F_1 = \overline{A}B + A\overline{B}$ gives another version of the exclusive-NOR gate as shown in Fig. 4.42.

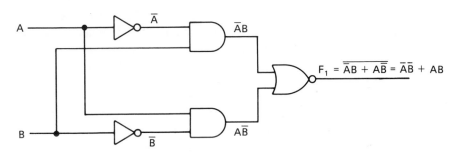

Fig 4.42 Alternative XNOR gate

4.12 One-Bit Magnitude Comparator

An example of a useful application of the exclusive-NOR circuit is a **magnitude comparator**. This is a circuit often used in process control and microprocessor systems to compare two one-bit numbers, A and B, and produce an output when $A = B$, $A < B$ or $A > B$. Therefore the magnitude comparator circuit for two variables will have two inputs and three outputs, of which only one output at a time will assume a logic level 1.

Let the three outputs be F_1, F_2 and F_3. Then

$F_1 = 1$ when $A = B = 0$ or $A = B = 1$, i.e. $A \equiv B$
$F_2 = 1$ when $A = 0$ and $B = 1$, i.e. $A < B$
$F_3 = 1$ when $A = 1$ and $B = 0$, i.e. $A > B$

The truth table for the specified conditions is shown in Table 4.13.

Table 4.13 Truth table for one-bit comparator

A	B	F_1	F_2	F_3
0	0	1	0	0
0	1	0	1	0
1	0	0	0	1
1	1	1	0	0

The output function for F_1 is $\overline{A}\,\overline{B} + AB$, which can be implemented by the XNOR gate of Fig. 4.38. The remaining two outputs are given by $F_2 = \overline{A}B$, and $F_3 = A\overline{B}$, which can be implemented as shown in Fig. 4.43.

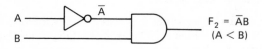

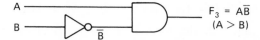

Fig 4.43 Implementation of F_2 and F_3

The three independent circuits for F_1, F_2 and F_3 can now be combined to form a one-bit comparator, as shown in Fig. 4.44.

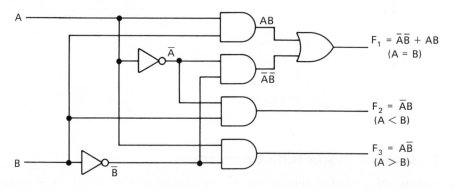

Fig 4.44 One-bit comparator

It must be pointed out that using the second version of the XNOR gate, Fig. 4.42 instead of Fig. 4.38, considerably simplifies the one-bit comparator, because no additional gates are required.

The XNOR gate, as shown in Fig. 4.45, has all three outputs readily available.

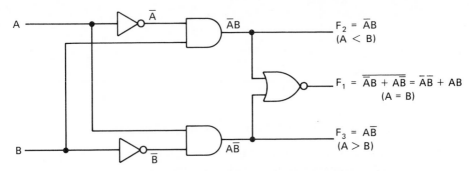

Fig 4.45 Simplified version of one bit comparator

4.13 Problems

4.1 Implement the following expressions using AND, OR and NOT gates.
(a) $F = A(\overline{A} + B)$ (b) $F = AC + AB + \overline{BC}$ (c) $F + \overline{A}C + \overline{B}C$
(d) $F = (A + B)(B + C)$

4.2 Implement the expressions given in Problem 4.1 using (a) NAND gates only and (b) NOR gates only.

4.3 Implement the XOR operation using NAND gates only.

4.4 Repeat Problem 4.3 but using NOR gates only.

4.5 Express the function of the following truth table (4.14) as a sum of products. Simplify the fundtion using Boolean algebra, and implement the function using AND, OR and NOT gates.

Table 4.14 Truth table for Problem 4.5

A	B	C	F
0	0	0	0
0	0	1	0
0	1	0	1
0	1	1	1
1	0	0	0
1	0	1	1
1	1	0	0
1	1	1	1

4.6 Repeat the implementation of Problem 4.5 using NAND gates only.

4.7 Repeat the implementation of Problem 4.5 using NOR gates only.

4.8 Derive the truth table and a Boolean expression for the circuit shown in Fig. 4.46.

4.9 Implement the expression derived in Problem 4.8 using AND, OR and NOT gates.

4.10 Repeat the implementation of Problem 4.8 using NAND gates only.

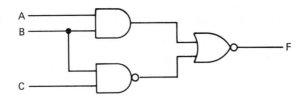

Fig 4.46   Diagram for Problem 4.8

4.11   Repeat the implementation of Problem 4.8 using NOR gates only.

4.12   Prove that for the conincidence gate, $A \odot B = \overline{A \oplus B}$, and show that the gate can be implemented with an exclusive-OR gate and an inverter.

4.13   Implement the coincidence gate using AND, OR and NOT gates.

4.14   Show that $F = A \oplus B \oplus C = A \oplus (B \oplus C) = (A \oplus B) \oplus C$ and implement the function using AND, OR and NOT gates.

4.15   Repeat the implementation of Problem 4.14 using NAND gates only.

4.16   Repeat the implementation of Problem 4.14 using NOR gates only.

4.17   Show that $F = A \odot B \odot C = A \odot (B \odot C) = (A \odot B) \odot C$ and implement the function using AND, OR and NOT gates.

4.18   Repeat the implementation of Problem 4.17 using NAND gates only.

4.19   Repeat the implementation of Problem 4.17 suing NOR gates only.

4.20   Show that if the 0's and 1's in the truth table of an AND gate are interchanged, it becomes a truth table for an OR gate.

4.21   Show that if the 0's and 1's in the truth table of an OR gate are interchanged, it becomes an AND gate.

4.22   Find a Boolean expression for the circuit shown in Fig. 4.47, then simplify the function and implement it.

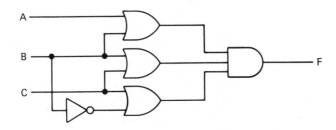

Fig 4.47   Logic circuit for Problem 4.22

4.23   Derive the truth table and a Boolean expression for the logic circuit shown in Fig. 4.48.

4.24   Using the truth table derived for Problem 4.23, show that F is given by $F = \overline{A}B + A\overline{C} + \overline{B}C$, and implement the function using AND, OR and NOT gates.

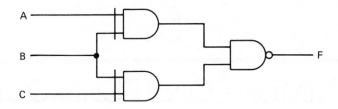

Fig 4.48   Logic circuit for Problem 4.23

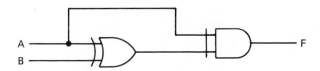

Fig 4.49   Logic circuit for Problem 4.25

4.25   Find F for the logic diagram shown in Fig. 4.49, then construct a truth table to show that F = B̄.

5

MINIMISATION TECHNIQUES

5.1 Introduction

There are three different methods commonly used for the minimisation of Boolean algebra expressions. These are **algebraic**, **mapping** and **tabular** methods. All these methods aim at reducing an original expression into a form with the lowest number of variables, and the lowest number of terms, in order to implement the function with the least number of gates, or the least number of integrated circuits. Less complexity means greater reliability and less expense.

To some extent, algebraic minimisation has been employed in 3 and 4 Chapters. To avoid repetition, algebraic minimisation will be briefly summarised, then the other two methods of minimisation will be discussed in detail.

5.2 Algebraic minimisation techniques

Methods of algebraic minimisation are as follows.

(i)   The rearrangement of algebraic expressions using factorisation.
(ii) The addition of zero terms.
(iii) Multiplication by a 1.
(iv) The expansion to canonical form.

Factorisation

Factorisation involves one of the following.

(a) The collection of terms having the same variable or a similar group of variables.
(b) A rearrangement of the terms by extracting the common factor, leaving a multiplier which itself reduces to a 1.

For example, $F = \overline{A}C + \overline{A}BC$ has a common factor of $\overline{A}C$, therefore it can be expressed as

$F = \overline{A}C(1 + B) = \overline{A}C$, since $(1 + B) = 1$.

Addition of zero terms

The addition of zero terms is often used in the transposition of Boolean algebra expressions for alternate implementation with different type of gates. This does not necessarily mean a reduction in the number of gates.

Multiplication by 1

Terms can be multiplied by a 1, using one of the following factors:

$$(1 + A) \qquad (A + \overline{A})$$
$$(1 + B) \qquad (B + \overline{B})$$

A simple but convincing example to illustrate this, is a function $F = A + \overline{A}B$ or $F = B + A\overline{B}$. Both of these functions can be simplified to $F = A + B$. This is done by multiplication of A by $(1 + B)$, or multiplication of B by $(1 + A)$ to obtain $A + B$. Alternatively, the expected answer $(A + B)$ can be multiplied by $(A + \overline{A})$ to obtain $F = A + \overline{A}B$, or it can be multiplied by $(B + \overline{B})$ to obtain $F = B + A\overline{B}$. This proves the identity. For a more complete solution, refer to Chapter 3 Example 3.8, but as a further proof the identity of these functions can be verified by constructing a truth table, as shown in Table 5.1.

Table 5.1   Truth table for $F = A + B = A + \overline{A}B = B + A\overline{B}$

A	B	F = A + B	$\overline{A}$	$\overline{A}B$	$F = A + \overline{A}B$	$\overline{B}$	$A\overline{B}$	$F = B + A\overline{B}$
0	0	0	1	0	0	1	0	0
0	1	1	1	1	1	0	0	1
1	0	1	0	0	1	1	1	1
1	1	1	0	0	1	0	0	1

Canonical form

Expansion to canonical form is often used in algebraic minimisation in order to eliminate redundant terms. For example, the expression $F = AB + AC + B\overline{C}$ can not be simplified directly, but expanding to the canonical form we get

$$F = AB(C + \overline{C}) + AC(B + \overline{B}) + B\overline{C}(A + \overline{A})$$
$$= ABC + AB\overline{C} + ABC + A\overline{B}C + AB\overline{C} + \overline{A}B\overline{C}$$

which reduces to $F = ABC + AB\overline{C} + A\overline{B}C + \overline{A}B\overline{C}$. This expression, in canonical form, can now be simplified by the factorisation method, combining the first term with the third and the second term with the fourth. This gives

$$F = AC(B + \overline{B}) + B\overline{C}(A + \overline{A}) = AC + B\overline{C}$$

which shows that the term AB of the original expression is redundant and therefore can be eliminated.

5.3   Veitch diagram and Karnaugh map

A logical function, whether in algebraic form or in the form of a truth table, can be represented diagrammatically using either a **Veitch diagram** or a **Karnaugh map**. Both methods use rectangular boxes subdivided into squares, sometimes referred to as **cells**. In the Veitch diagram, each cell accommodates one minterm of an algebraic expression in a canonical form; in a Karnaugh map, it holds a 0 or a 1 from a truth table, representing the expression.

Veitch diagrams and Karnaugh maps are both drawn in sizes with 2^n squares, starting from $n = 2$ for a two-variable function up to $n = 6$ for a six-variable function. The diagrammatic method for larger number of variables become too difficult to handle, and so alternate means are employed, as shown in Section 5.5.

Veitch diagrams and Karnaugh maps are very similar; the basic difference is in the way

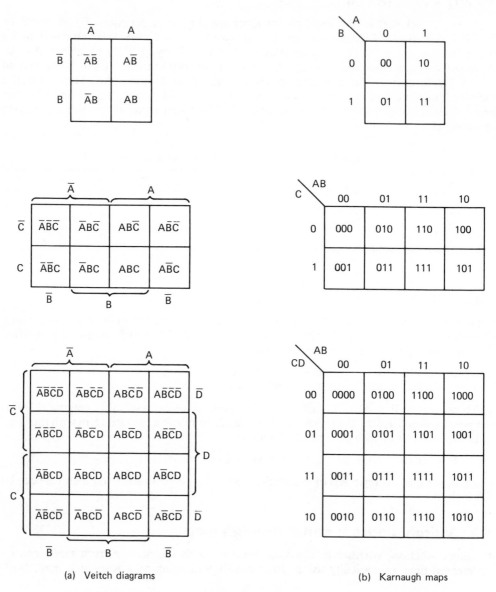

(a)   Veitch diagrams

(b)   Karnaugh maps

Fig 5.1   Pictonal representations of two, three and four variable functions

the columns and the rows are labelled. Figure 5.1 illustrates this for two, three and four variables with Veitch diagrams in Fig. 5.1(a) and Karnaugh maps in Fig. 5.1(b).

The common feature of these methods is that the two adjoining cells, both vertically and horizontally, differ in one variable only.

The individual cells on Veitch diagrams are marked as shown in Fig. 5.1(a), where the literals are in alphabetic order. On Karnaugh maps, Fig. 5.1(b), the variables are marked on each side of a diagonal drawn upwards from the top left hand corner of the map. In the simplest case, variable A is allocated to columns and B to rows. In the three-variable case, the variables A and B, are allocated in alphabetical order to the columns; and the variables below the diagonal, C and D, are allocated to the rows.

Each cell of a Veitch diagram represents the literal form of each term of a function covering all possible variations of n variables, whereas each cell of a Karnaugh map displays a binary form of each term. Therefore, if a function is expressed as a sum of T terms,

$$F = T_0 + T_1 + T_2 + \text{---} \quad + T_{n-1} = \sum_{k=0}^{n-1} T_k$$

where each cell of the Karnaugh map is identified with a number corresponding to the suffix of the term. Thus a logic function represented in a form of a truth table can be directly transferred, or mapped, onto the Karnaugh map. If the function is given as an algebraic expression it is initially easier to transfer it onto a Veitch diagram.

The cells are left blank prior to mapping. Only the available minterms of an expression, or the 1's of a truth table, representing a function are plotted onto the diagrams for simplification purposes. Missing term, or zeros of the truth table, are marked with 0's in the appropriate cells or are left blank.

The terminology relating to mapping will now be explained. Adjacency is a word used to indicate two plotted terms in adjoining cells, either horizontally or vertically but not diagonally. This is shown in Fig. 5.2, where the outside labelling of the maps is omitted for clarity.

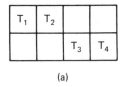

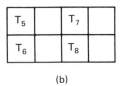

(a)   (b)

Fig 5.2   Adjacencies (a) between terms in a row; (b) between terms in a column

Thus in Fig. 5.2(a) the term T_1 is adjacent to T_2, T_3 is adjacent to T_4, but T_2 is not adjacent to T_3. This indicates adjacencies between terms in a row. Similarly, adjacencies between terms in a column are shown in Fig. 5.2(b), where T_5 is adjacent to T_6, and T_7 is adjacent to T_8. Adjacencies can also occur between

(a) rows externally

(b) columns externally

(c) four corner cells (a combination of rows and columns)

This is shown in Fig. 5.3

As mentioned earlier, Veitch diagrams are used for Boolean algebra expressions in literal form, whereas Karnaugh maps are employed for functions represented by a truth table. If a

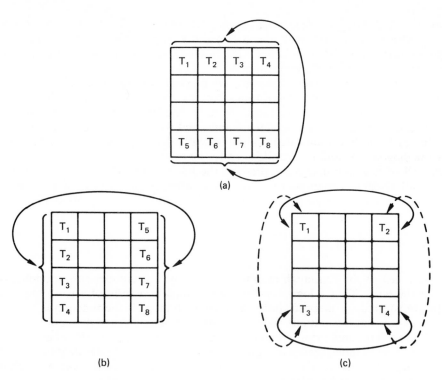

Fig 5.3   (a) Row; (b) column; (c) corner adjacencies

circuit is to be designed, the first stage is to produce a truth table. This can be plotted directly onto a Karnaugh map without having to derive the function. For this reason Karnaugh maps will be used in preference to Veitch diagrams. The peripheral labelling of rows and columns, however, will be combined using both Veitch and Karnaugh methods. This will be done to clarify the order in which the variables are used in the truth table and the way in which the terms are transferred onto the Karnaugh maps.

Truth tables and Karnaugh maps can be presented in two different ways:

(i)   with A as the LSD (least significant digit) or LSB (least significant bit);
(ii)   with A as the MSD (most significant digit).

This is shown in Table 5.2 and Figures 5.4 and 5.5, for two, three and four variables.

The advantage of using A as LSD is that one truth table can cover two-. three- and four-variable Karnaugh maps, and the truth table can easily be expanded to cover more variables. Additional variables can be inserted on the left hand side of the truth table without having to rewrite the whole truth table. This is shown in Table 5.2(c), where F_c is a four-variable, F_b is a three-variable and F_a is a two-variable function; all three functions are displayed on the same truth table. A further advantage of using A as the LSD will become apparent in Chapters 8 and 9 where, in the case of registers and counters, literals in alphabetical order correspond to both the binary count and the binary content of the cell, in the case of pure binary codes and binary coded decimal codes.

Throughout this chapter, when Karnaugh maps are referred to, the decimal numbering of the cells refers to the binary equivalent of the content of the cell. This is also used in Chapter 9, where Karnaugh maps are used for counter design.

Table 5.2 (i) A = LSD in (a), (b) and (c)
(ii) A = MSD in (d), (e) and (f)

(a)		B	A	F_a		(d)		A	B	F_d
	0	0	0	T_0			0	0	0	T_0
	1	0	1	T_1			1	0	1	T_1
	2	1	0	T_2			2	1	0	T_2
	3	1	1	T_3			3	1	1	T_3

(b)		C	B	A	F_b		(e)		A	B	C	F_e
	0	0	0	0	T_0			0	0	0	0	T_0
	1	0	0	1	T_1			1	0	0	1	T_1
	2	0	1	0	T_2			2	0	1	0	T_2
	3	0	1	1	T_3			3	0	1	1	T_3
	4	1	0	0	T_4			4	1	0	0	T_4
	5	1	0	1	T_5			5	1	0	1	T_5
	6	1	1	0	T_6			6	1	1	0	T_6
	7	1	1	1	T_7			7	1	1	1	T_7

(c)		D	C	B	A	F_c	F_b	F_a		(f)		A	B	C	D	F_f
	0	0	0	0	0	T_0	T_0	T_0			0	0	0	0	0	T_0
	1	0	0	0	1	T_1	T_1	T_1			1	0	0	0	1	T_1
	2	0	0	1	0	T_2	T_2	T_2			2	0	0	1	0	T_2
	3	0	0	1	1	T_3	T_3	T_3			3	0	0	1	1	T_3
	4	0	1	0	0	T_4	T_4				4	0	1	0	0	T_4
	5	0	1	0	1	T_5	T_5				5	0	1	0	1	T_5
	6	0	1	1	0	T_6	T_6				6	0	1	1	0	T_6
	7	0	1	1	1	T_7	T_7				7	0	1	1	1	T_7
	8	1	0	0	0	T_8					8	1	0	0	0	T_8
	9	1	0	0	1	T_9					9	1	0	0	1	T_9
	10	1	0	1	0	T_{10}					10	1	0	1	0	T_{10}
	11	1	0	1	1	T_{11}					11	1	0	1	1	T_{11}
	12	1	1	0	0	T_{12}					12	1	1	0	0	T_{12}
	13	1	1	0	1	T_{13}					13	1	1	0	1	T_{13}
	14	1	1	1	0	T_{14}					14	1	1	1	0	T_{14}
	15	1	1	1	1	T_{15}					15	1	1	1	1	T_{15}

5.4 Simplification of Boolean functions using Karnaugh maps

Mapping and demapping of two-variable functions

Mapping is a technique of representing a Boolean function on a Karnaugh map. This means placing a 1 in a square which indicates the presence of that term in the given function.

Example 5.1
Map the two-variable function $F = \bar{A}\bar{B} + A\bar{B}$ and simplify, if possible.

Solution
The function given in Example 5.1 is plotted in Fig. 5.6.

It can be seen that an adjacency exists between two terms. The adjacency is enclosed in a loop which contains a common factor $\bar{B}$. In fact, the adjacency is $\bar{B}$, i.e. the function $F = \bar{B}$ because $\bar{A}\bar{B} = \bar{B}(\bar{A} + A) = \bar{B}$, since $A + \bar{A} = 1$.

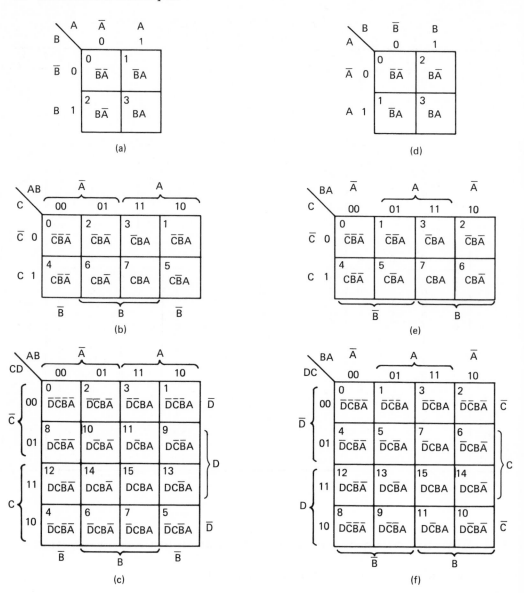

Fig 5.4   Karnaugh maps with A as least significant digit

Looping, often referred to as **grouping** or forming subcubes, uncovers common factors. This leads to a simplified expression, since at least one of the factors will reduce to a 1. What remains to be done is to write a final or simplified version of the original function. This is called **demapping,** and it involves extracting the minimised form of the function directly from the Karnaugh map without resorting to Boolean algebra minimisation. The looped adjacencies are refered to as **prime implicants.**

Looping adjacent terms and demapping will further be pursued with three- and four-

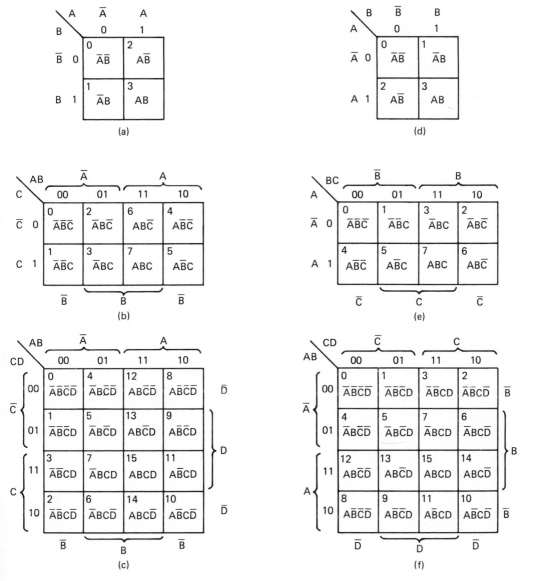

Fig 5.5 Karnaugh maps with A as most significant digit

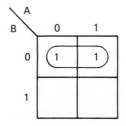

Fig 5.6 Karnaugh maps for $F = \overline{A}\overline{B} + A\overline{B}$

variable maps, but first more examples of two-variable functions will be examined in order to explain some additional properties of Karnaugh maps.

Example 5.2
Simplify the function $F = \overline{A} + \overline{A}B$ using a Karnaugh map.

Solution
In this example the first term occupies both cells in column $\overline{A}$. The second term is redundant, since it also occupies the bottom left hand cell of the map.

To plot the function, ensure that it is in canonical form, i.e. the sum-of-product form, in which every term contains all the variables of the given function. That is: $F = \overline{A} + \overline{A}B = \overline{A}(B + \overline{B}) + \overline{A}B = \overline{A}B + \overline{A}\overline{B} + \overline{A}B$. Therefore, $F = \overline{A}B + \overline{A}\overline{B}$ which is mapped as shown in Fig. 5.7.

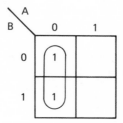

Fig 5.7   Karnaugh map for $F = \overline{A} + \overline{A}B$

Looping the adjacent terms and demapping gives $F = \overline{A}$. It is obvious if a reference is made to Boolean algebra, but similar examples may not be so obvious when more variables are introduced.

Example 5.3
Map the function $F = \overline{A}B + A\overline{B}$ and explain why simplification is not possible.

Solution
The given function is already in the canonical form, and therefore it can be directly mapped as shown in Fig. 5.8.

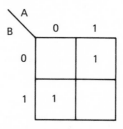

Fig 5.8   Karnaugh map for $F = \overline{A}B + A\overline{B}$

There is no adjacency in this case, and therefore no looping is possible, showing that the given function is made up of essential prime implicants only, that is terms which cannot be looped. The simplest case of this is illustrated by Example 5.4 which includes overlapping loops.

Example 5.4
Simplify by Karnaugh mapping and check by an algebraic method the function
$F = \overline{A}\overline{B} + A\overline{B} + AB$.

Solution
(i) The Karnaugh map as shown in Fig. 5.9.

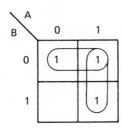

Fig 5.9   Karnaugh map for $F = \overline{A}\overline{B} + A\overline{B} + AB$

Demapping gives $F = A + \overline{B}$, since there are two adjacencies with overlapping loops. Overlapping loops are allowed, and each loop is demapped separately.

(ii) By Boolean algebra,

$$
\begin{aligned}
F &= \overline{A}\overline{B} + A\overline{B} + AB \\
  &= \overline{B}(\overline{A} + A) + AB \\
  &= \overline{B} + AB \\
  &= \overline{B} + A.
\end{aligned}
$$

The simple example 5.4 convincingly illustrates the advantage of the Karnaugh map method over the algebraic method, since the minimal function is directly extracted from the map. If all four cells are 1's, for a two-variable function $F = 1$. To determine the minimal logic for $\overline{F}$ the cells containing 0's are looped by the same method. Thus in the Example 5.2, if the unmarked cells are taken as 0's, then $\overline{F} = A\overline{B} + AB$. Looping the two adjacent terms and demapping gives $\overline{F} = A$, which checks with the solution of the Example 5.2, since $F = \overline{A}$.

Mapping and demapping of three-variable functions

For a three-variable function, the Karnaugh map has $2^n = 2^3 = 8$ squares or cells. The mapping procedure is the same as for two-variable functions, i.e. all minterms of a given function are represented by 1's in the appropriate squares and the adjacent 1's are looped in corresponding powers of 2.
    Demapping a three-variable function as follows.

(i)   all eight terms gives $F = 1$
(ii)   four terms gives F in the form of one variable only
(iii)   two terms gives F in the form of a two-variable term
(iv)   one term gives F in the form of a three-variable term

Looping must continue until all adjacencies, both internally and externally, are exhausted, allowing for overlapping of terms.

Example 5.5
Map and simplify the function $F = \overline{A}\overline{B}C + A\overline{B}C + A\overline{B}\overline{C}$.

Solution

The function is mapped as shown in Fig. 5.10.

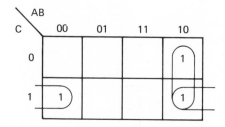

Fig 5.10   Karnaugh map for $F = \overline{A}B\overline{C} + A\overline{B}C + A\overline{B}\overline{C}$

There are two adjacencies, as indicated by the two loops: one between terms in the right hand column and the other between corner terms in the bottom row, with one term overlapping.

Demapping the function gives $F = A\overline{B} = \overline{B}C$.

Example 5.6

Simplify, using a Karnaugh map, the function $F = \overline{A}\overline{B}\overline{C} + \overline{A}B + AB + A\overline{B}\overline{C}$.

Solution

This time the function has to be put in canonical form first. This gives

$$F = \overline{A}\overline{B}\overline{C} + \overline{A}B + AB + A\overline{B}\overline{C} = \overline{A}\overline{B}\overline{C} + \overline{A}B(\overline{C} + C) + AB(\overline{C} + C) + A\overline{B}\overline{C}$$
$$= \overline{A}\overline{B}\overline{C} + \overline{A}B\overline{C} + \overline{A}BC + AB\overline{C} + ABC + A\overline{B}\overline{C}$$

There are two loops of four terms each, allowing for overlapping as shown in Fig. 5.11. Thus $F = B + \overline{C}$.

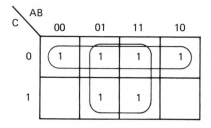

Fig 5.11   Karnaugh map for Example 5.6

Four-variable functions

Karnaugh mapping for four-variable functions follows the same guidelines as for two and three variables. Thus, there are $2^4 = 16$ squares in the Karnaugh map for four variables. The variables are divided into two groups, say AB and CD, and the diagram is labelled as shown in Fig. 5.4 or Fig. 5.5.

When demapping a four-variable function, the minimal logic expression for F will result from the following sequential groups:

(a) sixteen 1's give F = 1
(b) eight 1's give   F = single variable term
(c) four 1's give   F = double variable term
(d) two 1's give   F = three variable term
(e) one bit 1 gives F = four variabe term

Grouping continues until all loops are considered, allowing for overlapping.

Example 5.7
Simplify the following functions using Karnaugh maps:

(a) $F = \bar{A}\bar{B}CD + \bar{A}BC\bar{D} + AB\bar{C}\bar{D} + \bar{A}BCD + ABCD + A\bar{B}C\bar{D} + \bar{A}BC\bar{D} + ABC\bar{D}$
(b) $F = \bar{A}B\bar{C}\bar{D} + \bar{A}\bar{B}C\bar{D} + AB\bar{C}\bar{D} + A\bar{B}\bar{C}D + \bar{A}\bar{B}\bar{C}D + A\bar{B}\bar{C}D + \bar{A}B\bar{C}D$
$\quad + A\bar{B}CD + \bar{A}\bar{B}C\bar{D} + \bar{A}BC\bar{D} + ABC\bar{D} + A\bar{B}C\bar{D} + \bar{A}\bar{B}C\bar{D}$
(c) $F = \bar{A}B\bar{C}\bar{D} + AB\bar{C}\bar{D} + \bar{A}BC\bar{D} + AB\bar{C}\bar{D}$

Solution
(a) Figure 5.12 shows that there are two loops of four 1's, which when demapped give
$F = B\bar{D} + CD$.

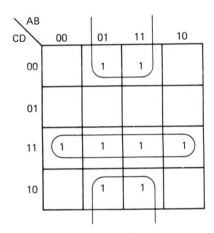

Fig 5.12   Karnaugh map for Example 5.7(a)

(b) This time there are two loops of eight 1's, reducing the function to two single-variable terms obtained from Fig. 5.13. This gives $F = \bar{B} + \bar{D}$.
(c) Here the four corner 1's form an outside loop giving one two-variable term $F = \bar{B}\bar{D}$, as shown in Fig. 5.14.

Karnaugh maps for inverse functions

All the functions mapped so far are of the sum-of-product form, which after simplification could be implemented by AND and OR gates. The inverse functions will be considered now. These are represented on the Karnaugh maps by 0's. In the algebraic form, inverse

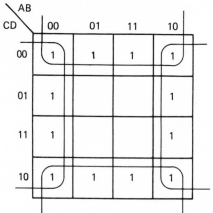

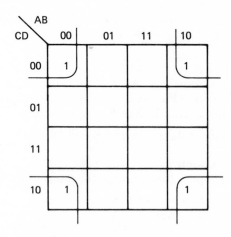

Fig 5.13 Karnaugh map for Example 5.7(b) **Fig 5.14** Karnaugh map for Example 5.7(c)

functions are represented as product-of-sums, which cannot directly be plotted on Karnaugh maps.

A function in the product-of-sum form has to be expressed in the sum-of-product form before it can be plotted.

Mapping the inverse function instead of the function itself often results in a simpler implementation of the function.

Example 5.8
In the solution of Example 5.5, the unmarked cells represented the inverse function, and would normally be filled in with 0's. Write the five terms of the inverse function, then plot them on a Karnaugh map and simplify.

Solution
The inverse function of Example 5.5 is $\bar{F} = \bar{A}\bar{B}\bar{C} + \bar{A}B\bar{C} + AB\bar{C} + \bar{A}BC + ABC$, which is plotted in Fig. 5.15.

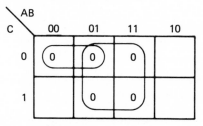

Fig 5.15 Karnaugh map for Example 5.8

Looping 0's and demapping gives $\bar{F} = B + \bar{A}\bar{C}$. Inverting the result to obtain the function gives $F = \overline{B + \bar{A}\bar{C}} = \bar{B}.\overline{\bar{A}\bar{C}} = \bar{B}(A + C)$, which can be implemented with three gates, whereas the original solution to Example 5.5 required four gates.

Another, perhaps more convincing, example will help to show that sometimes it is better to use Karnaugh map for the inverse function and then invert the result before implementing the solution. This technique is considered next.

Example 5.9

Plot, simplify and implement the function $F = \overline{A}\overline{B}\overline{C}\overline{D} + \overline{A}B\overline{C}\overline{D} + \overline{A}\overline{B}\overline{C}D + AB\overline{C}\overline{D} + A\overline{B}\overline{C}D + A\overline{B}CD$. Then plot the inverse function, simplify it, and implement it in order to compare the complexity of the implemented circuits.

Solution

The function F is mapped in Fig. 5.16.

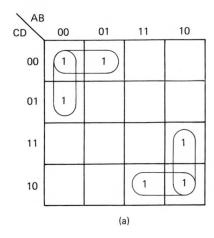

(a)

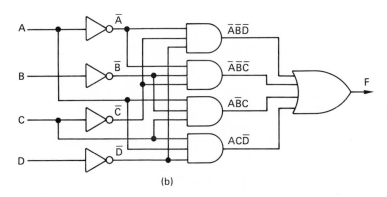

(b)

Fig 5.16   (a) Karnaugh map for the function; (b) its implementation

The simplified solution of the function is $F = \overline{A}C\overline{D} + \overline{A}\overline{B}\overline{C} + A\overline{B}C + AC\overline{D}$, which is implemented in Figure5.16(b). To implement this function four inverters, four three-input AND gates and an four-input OR gate are required. In addition, 22 interconnections are needed.

Now, the unmarked spaces on the Karnaugh map of Fig. 5.16 represent the inverse function, $\overline{F}$. Plotting the inverse function separately in Fig. 5.17(a) and demapping gives $\overline{F} = A\overline{C} + BD + \overline{A}C$. To generate the function from the inverse, an additional inverter is required, as shown in Fig. 5.17(b). However, even then only three inverters, three two-input AND gates and a three-input OR gates are required, with only 13 interconnections.

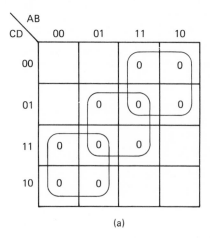

(a)

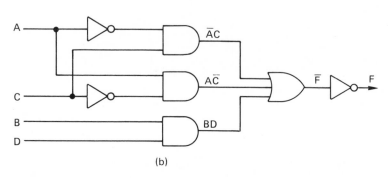

(b)

Fig 5.17   (a) Karnaugh map for the inverse function; (b) and its implementation

The two implementations clearly indicate that the inverse function provides a simpler solution, with fewer gates, and with simpler gates which require fewer interconnections. The rule is to identify whether there are more 1's than 0's, or vice-versa. If there are more 1's, use the direct method; if there are more 0's use the inverse method.

Unused terms, forbidden terms, redundant terms or don't care terms

All four names apply to similar situations. They describe combinations of variables which never occur in practice.

In general, the redundant terms are not available at all, or are rejected, or are purposely suppressed so that in practice they can not happen and therefore can be used for simplification purposes. The terms are marked X, x, R or d, on the Karnaugh map, and they can be treated as 1's or 0's for adjacencies and looping. Standard examples quoted usually refer to the binary coded decimal (BCD) codes, where four binary digits are used to represent the ten decimal digits. Since there are sixteen possible binary combinations, six of the available terms are redundant.

BCD codes and the design of code converters are dealt with elsewhere. Here Karnaugh maps are used purely for simplification purposes of algebraic functions.

In most cases, A will be treated as the least significant bit or least significant binary digit, abbreviated to LSB or LSD. This is shown in Table 5.3 and Fig. 5.18.

Table 5.3 BCD truth table

Decimal value	Binary equivalent				Implementation				Simplified implementation			
	D	C	B	A	D	C	B	A	D	C	B	A
0	0	0	0	0	$\bar{D}$	$\bar{C}$	$\bar{B}$	$\bar{A}$	$\bar{D}$	$\bar{C}$	$\bar{B}$	$\bar{A}$
1	0	0	0	1	$\bar{D}$	$\bar{C}$	$\bar{B}$	A	$\bar{D}$	$\bar{C}$	$\bar{B}$	A
2	0	0	1	0	$\bar{D}$	$\bar{C}$	B	$\bar{A}$		$\bar{C}$	B	$\bar{A}$
3	0	0	1	1	$\bar{D}$	$\bar{C}$	B	A		$\bar{C}$	B	A
4	0	1	0	0	$\bar{D}$	C	$\bar{B}$	$\bar{A}$		C	$\bar{B}$	$\bar{A}$
5	0	1	0	1	$\bar{D}$	C	$\bar{B}$	A		C	$\bar{B}$	A
6	0	1	1	0	$\bar{D}$	C	B	$\bar{A}$		C	B	$\bar{A}$
7	0	1	1	1	$\bar{D}$	C	B	A		C	B	A
8	1	0	0	0	D	$\bar{C}$	$\bar{B}$	$\bar{A}$	D			$\bar{A}$
9	1	0	0	1	D	$\bar{C}$	$\bar{B}$	A	D			A
10	1	0	1	0			X				X	
11	1	0	1	1			X				X	
12	1	1	0	0			X				X	
13	1	1	0	1			X				X	
14	1	1	1	0			X				X	
15	1	1	1	1			X				X	

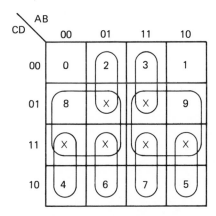

Fig 5.18 Karnaugh map for BCD logic

Redundant or don't care terms can be taken as 1's if this helps with the simplification, or else they can be taken as 0's and ignored.

5.5 Tabular minimisation

The minimisation of logic functions by a tabular method, introduced by Quine and McCluskey, is usually used for functions with more than four variables. The method involves matching terms which differ by only one digit. To illustrate the process the tabular minimisation of a four variable function will be given as a worked example.

Example 5.10
Simplify the function:

$$F = \overline{A}\overline{B}\overline{C}D + \overline{A}\overline{B}CD + AB\overline{C}\overline{D} + AB\overline{C}D + ABCD + ABC\overline{D} + A\overline{B}CD + A\overline{B}\overline{C}D$$

Solution

Step 1. The function to be minimised is written out in canonical form. In this case step 1 can be omitted.

Step 2. The terms of the function are arranged in tabular form, starting with the minimum number of uncomplemented variables, (i.e. minimum number or 1s), when written out in an equivalent binary form. This is given in Table 5.4, columns 3 and 4.

Table 5.4

			(a)
No. of 1's	Decimal equivalent	Canonical form	Binary equivalent
1	1	$\overline{A}\overline{B}\overline{C}D$	0001 ✔
2	3	$\overline{A}\overline{B}CD$	0011 ✔
	9	$A\overline{B}\overline{C}D$	1001 ✔
	12	$AB\overline{C}\overline{D}$	1100 ✔
3	11	$A\overline{B}CD$	1011 ✔
	13	$AB\overline{C}D$	1101 ✔
	14	$ABC\overline{D}$	1110 ✔
4	15	$ABCD$	1111 ✔

Step 3. The terms are arranged in groups of single 1s, two 1s, three 1s, etc., as shown in rows 1 to 4 in Table 5.4.

Step 4. A column of decimal equivalents is set out for ease of identification. (See Table 5.4, column 2.)

Step 5. Matched terms are collected, marking each term with a tick (✔) and entering the simplified version into a new list. For instance $\overline{A}\overline{B}\overline{C}D$ and $\overline{A}\overline{B}CD$ have a common factor $\overline{A}\overline{B}D$ and so they simplify to $\overline{A}\overline{B}D$, since $(C + \overline{C}) = 1$. In the simplified list a dash (—) is substituted to signify $C + \overline{C}$, giving 00-1 in the binary column of Table 5.5.

Table 5.5

	(b)
Decimal equivalent	Binary match
1, 3	00–1 ✔
1, 9	–001 ✔
3, 11	–011 ✔
9, 11	10–1 ✔
9, 13	1–01 ✔
12, 13	110– ✔
12, 14	11–0 ✔
11, 15	1–11 ✔
13, 15	11–1 ✔
14, 15	111– ✔

Step 6. Each term in one group is matched with every term in the next group in column 4 of Table 5.4. The terms are ticked and are entered into the new list of Table 5.5 which now contains terms of 3 variables.

Step 7. The matching procedure is now applied to Table 5.5, producing Table 5.6 which contains only two variables. This procedure would be repeated (Step 8) to give further reduction to a single variable if possible. In this example no further reduction is possible and so Step (8) is omitted.

Table 5.6

(c)	
Decimal equivalent	Binary match
1, 3; 9, 11	−0−1 }P
1, 9; 3, 11	−0−1
9, 11; 13, 15	1−−1 }Q
9, 13; 11, 15	1−−1
12, 13; 14, 15	11−− }R
12, 14; 13, 15	11−−

Step 9. All lists are now inspected for unticked terms. These terms are not implied by any other terms and they are called **prime implicants**. In this example the prime implicants are denoted by P, Q, and R.

Step 10. Further minimisation may be performed by inspection of prime implicants. This is done with the help of a prime implicants table as in Table 5.7 which contains prime implicants P, Q, and R, and is used to find a minimal subset of prime implicants which will include, or cover, all the terms in the original function.

The prime implicant Q is covered by P and R, i.e. the decimal values corresponding to Q(9, 11, 13, 15) are included in P(9, 11) and R(13, 15), therefore the function F is simplified to F = P + R; and since P is represented by $\bar{B}D$ and R is given by AB, the function F = $\bar{B}D$ + AB.

Table 5.7

Prime implicants	Decimal equivalents							
	1	3	9	11	12	13	14	15
P −0−1	X	X	X	X				
Q 1−−1			X	X		X		X
R 11−−					X	X	X	X

Table 5.8   Matrix of prime implicants

		Decimal equivalents							
		1	3	9	11	12	13	14	15
P	1, 3, 9, 11	X	X	X	X				
Q	9, 11, 13, 15			X	X		X		X
R	12, 13, 14, 15					X	X	X	X

An alternative method, which is somewhat simpler, is to draw a grid (or matrix) of lines corresponding to the prime implicants forming rows and decimal equivalents represented by columns, as shown in Table 5.8. The corresponding intersections are then marked with X's.

The final result from Table 5.8 is the same as from Table 5.7, i.e. $F = \overline{B}D + AB$.

5.6 Problems

5.1 Minimise the following expressions using Karnaugh maps:-

(a) $\overline{A}B\overline{C} + ABC + \overline{A}C + A\overline{B}C$

(b) $A\overline{B} + AC + \overline{A}BC + BC$

(c) $\overline{A}BCD + \overline{A}\overline{B}C + \overline{A}C\overline{D} + B\overline{C}D + AB\overline{C}\overline{D}$

(d) $\overline{A}\overline{B}C + \overline{A}BCD + \overline{B}D + BD + \overline{A}D$

(e) $\overline{A}C\overline{D} + \overline{A}CD + A\overline{B}CD + \overline{A}B\overline{C}\overline{D} + AB\overline{C}D + \overline{A}\overline{B}C + \overline{A}B\overline{D}$

(f) $\overline{A}BCD + \overline{A}B\overline{C}D + \overline{A}B\overline{C}\overline{D} + ABCD + \overline{A}BC\overline{D} + AD$

(g) $\overline{A}\overline{B}\overline{C}\overline{D}\overline{E} + \overline{A}BCDE + ABDE + \overline{A}C\overline{D}\overline{E} + A\overline{B}\overline{C}D + \overline{B}CD\overline{E} + B\overline{C}DE$

(h) $ABCDEF + \overline{A}B\overline{C}D\overline{E}F + \overline{A}BCD\overline{E}F + AB\overline{C}D\overline{E}F + \overline{A}BCDEF + A\overline{B}\overline{C}D\overline{E}F + \overline{A}\overline{B}CDEF + ABCD\overline{E}\overline{F}$

(i) $AB\overline{C}D + \overline{A}BD + \overline{A}BCD + \overline{A}B\overline{C}\overline{D}$ (Don't care terms $ABCD$, $\overline{A}\overline{B}CD$, $\overline{A}BC\overline{D}$)

5.2 Simplify the function $F = \overline{A}B\overline{C}\overline{D} + \overline{A}B\overline{C}D + \overline{A}\overline{B}CD + \overline{A}BC\overline{D} + ABCD + \overline{A}BCD + A\overline{B}\overline{C}D + \overline{A}\overline{B}CD$ using the tabular method.

6

THE CHARACTERISTICS OF LOGIC FAMILIES

6.1 Introduction

It is important to have some knowledge of the internal design of logic gates, even though they are normally obtained in integrated circuit form. Gates must not be considered merely as generators of logical functions which will operate under all conditions – this is certainly not the case. Symbolic representation of gate circuits do tend to encourage this view. Logic diagrams using gate symbols give no indication of power supply requirements or details of any physical components, perhaps giving a false impression to the beginner that there are no restrictions on how the gates are interconnected as long as the circuit is logically correct. The budding logic circuit designer must therefore be aware of this and ensure that the electrical requirements of the design are satisfied as well as the logical requirements. Full manufacturer's details are always available on logic gates and provide information in the form of individual data sheets or in a comprehensive data book. Clearly it is essential to be able to understand and interpret this data to use the information it contains correctly and this chapter aims to provide a background of knowledge to allow this to be done. Having decided upon the particular logic function required the designer must consider questions such as:

(1) What delay between applying a particular input signal to each gate and its arrival at the output can be tolerated?
(2) What is the maximum power dissipation that is permitted?
(3) How many circuits are to be driven by each gate, and what is the total current requirement?
(4) What electrical noise level will each gate have to withstand?

If high speed, (i.e. high frequency) operation is required, then the delay caused by the gate must be small. If power dissipation must be low, then the particular circuit design must be such that only a small current is drawn from the power supply. If a large number of circuits are to be driven, then the type of gate chosen must be able to produce the required drive current. If the gate is to be used in an environment where there are large electrical noise levels, the gate must be able to tolerate induced noise without producing false outputs.

Logic gates are available in a number of different **logic families**. A logic family comprises a set of devices, manufactured using a particular process, that includes the commonly required functions. This will include devices other than gates, such as flip-flops, registers, memories and microprocessors. Each member of a particular logic family has the same basic characteristics that are peculiar to that logic family. This means that gates from one logic family can be used if, for example, low power consumption is the prime requisite. If high speed operation is the requirement, then a different logic family could be used. Until recently there was no possibility of combining high speed operation with low power consumption. Great strides have been taken in recent years, however, and it is now possible to obtain a satisfactory compromise for most applications. Some families are implemented predominantly in terms of the fairly simple functions (such as gates), while others are used mainly for the implementation of complex digital circuits such as microprocessors. In other words, it is not possible to find equivalent functions in all the logic families.

6.2 Logic families

The logic families to be considered in this chapter are as follows.

(1) Diode–transistor logic (DTL)
(2) Transistor–transistor logic (TTL)
(3) P-channel metal oxide semiconductor logic (PMOS)
(4) N-channel metal oxide semiconductor logic (NMOS)
(5) Complementary metal oxide semiconductor logic (CMOS)
(6) High speed CMOS (HCMOS)
(7) Emitter-coupled logic (ECL)
(8) Integrated injection logic (I²L)

DTL will be considered mainly from the viewpoint of an introduction to TTL. Although DTL is still to be found in some equipment, it has been superseded by TTL. TTL, which has been available since 1964, provides high speed logic at low cost. PMOS logic permits cheap, high component density for low speed applications. NMOS technology allows a higher density of components than PMOS, and is faster but costs more. CMOS logic is widely used because of its very low power consumption and is suitable for battery applications. High speed CMOS is a combination of the low power consumption of the conventional CMOS and the speed of TTL, most circuits having compatible electrical connections and the same functions as their TTL equivalents. ECL is the fastest logic available, at the expense of considerably increased power dissipation. I²L is a relatively undeveloped form of logic that is still to gain a hold on the market.

It may be possible to develop I²L so that it combined the best features of all the other technologies previously mentioned.

6.3 The bipolar transistor as a switch

A bipolar transistor can be used in such a way that it's collector voltage swings between two extremes. With a (theoretically) perfect device, this is exactly equivalent to a conventional switch, as illustrated in Fig. 6.1(a)) and (b).

With the circuit shown, an open switch will prevent all current flowing from the supply voltage V and there will be no potential drop across the load resistor R_L. The output voltage will therefore be the same as the supply voltage. When the switch is closed, the potential difference across the load resistor is $V = 0\ V$ volts, and a current $I = V/R_L$ flows through the load. The output is held at ground potential, so the output voltage $V_{out} = O\ V$. In this situation a logic 1 could be represented by voltage V (switch open) and a logic 0 by O V (switch closed).

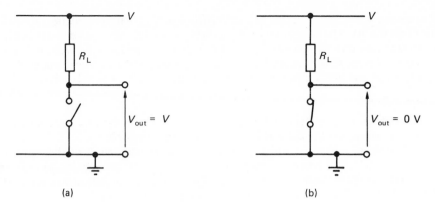

Fig 6.1   (a) open switch; (b) closed switch

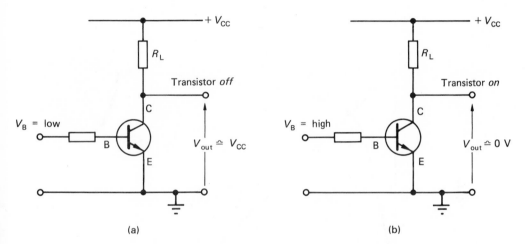

Fig 6.2   (a) Bipolar transistor OFF; (b) Bipolar transistor ON

To implement an NPN bipolar transistor as a switch, the circuit shown in Fig. 6.2 can be used.

In Fig. 6.2(a), the base–emitter junction is reverse biased by holding the base voltage at a suitably low potential with respect to the emitter. This prevents the transistor from conducting (in an exactly similar manner to the open switch in Fig. 6.1(a)), with the result that the output voltage will rise to a value close to V_{CC}. If the transistor could switch off completely, then of course the output voltage would be exactly equal to V_{CC}; but this can only happen if the impedance between the collector and emitter is infinite. At an impedence less than infinite, a current will flow from the supply through the load resistor and the transistor to ground.

With the situation illustrated in Fig. 6.2(a), the transistor is regarded as being *off*; but a small leakage current I_{CEO} will flow, causing a potential drop across the load resistance equal to $I_{CEO}R_{L}$. The output voltage will thus be given by:

$$V_{out} = V_{CC} - I_{CEO}R_{L}$$

In Fig. 6.2(b), the base–emitter junction is forward biased by holding the base voltage at a suitably high potential with respect to the emitter. The transistor will conduct heavily, and will saturate at a maximum value of collector current $I_{C(max)}$. This is equivalent to the closed switch of Fig. 6.1(b). The output voltage will drop to a value very close to zero volts. The output voltage can only be exactly equal to O V if the impedance between collector and emitter becomes zero, which would cause the entire supply voltage to be dropped across the load resistance R_L. If this happened, then the collector current would be given by:

$$I_C = V_{CC}/R_L$$

As the transistor is not a perfect switch, $I_{C(max)}$ will be less than V_{CC}/R_L and a small voltage $V_{CE(sat)}$ will remain across the collector – emitter of the transistor.

This is illustrated in Fig. 6.3 which gives the V_{CE}/I_C characteristic of the transistor for zero base current ($I_B = 0$) corresponding to an *off* transistor and maximum base current $I_{B(max)}$ corresponding to a saturated transistor. The load line which is drawn for particular values of collector supply voltage and load resistance is superimposed.

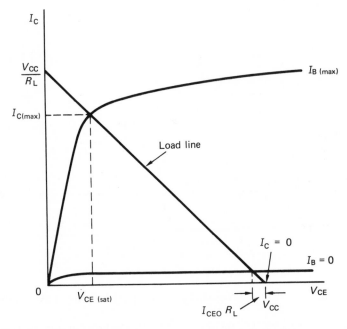

Fig 6.3   Transistor output characteristic with load line

The intersection of the load line and the appropriate characteristic gives the relevant values of collector current and collector–emitter voltage. With $I_B = 0$ it can be seen that a voltage $I_{CEO}R_L$ is developed across the load resistance. When $I_C = I_{C(max)}$ the output voltage is $V_{CE(sat)}$. The voltage output therefore will swing between $V_{CE(sat)}$ and $V_{CC} - I_{CEO}R_L$.

It can be seen from Fig. 6.2(a) and (b) that when a high input is applied to the transistor a low output results, and when a low input is applied a high output is obtained. The transistor acts as an **inverter** or **NOT gate**, converting a logic 1 to a logic 0 and vice versa. Connecting two NOT gates in series would change a 1 to a 0 and back to a 1 again, without degrading the voltage level, because of the inherent gain of the transistors. This feature is important, in the inverter considered and in all logic gates involving an output transistor.

Logic gates that operate on the principle of swinging the collector voltage between cut-off and saturation are classified as **saturating logic**. An example of saturating logic is TTL.

The disadvantage of saturating logic is that the maximum frequency of operation is limited because of the delay in switching off a transistor that is deep in saturation. The delay occurs because excess charge is stored in the base region, and has to be removed before the transistor can start to switch off.

The **propagation delay** in a stage is defined in Fig. 6.4 which uses a NAND gate connected as an inverter.

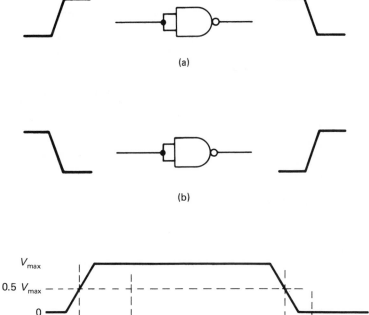

(a)

(b)

(c)

Fig 6.4   (a) Effect of a low to high transition on a NOT gate; (b) Effect of a high to low transition on a NOT gate; (c) Timing delays in a NOT gate

Fig. 6.4(a) shows the effect of applying a low to high transition to the gate, Fig. 6.4(b) the effect of applying a high to low transition, and Fig. 6.4(c) the timing delays involved. The overall propagation delay is given by averaging the values of t_{pHL} and t_{pLH}:

$$I_p = (t_{pHL} + t_{pLH})/2$$

Because of charge storage, the delay in a low to high transition at the output of such a gate would be greater than that for a high to low transition.

Table 6.1 Switching characteristics at $V_{CC} = 5$ V, $T_A = 25°C$. (Courtesy of Texas Instruments Inc.)

Type	Test conditions[#]	t_{pLH} (ns) Propagation delay time, low-to-high-level output			t_{pHL} (ns) Propagation delay time, high-to-low-level output		
		MIN	TYP	MAX	MIN	TYP	MAX
'00, '10			11	22		7	15
'04, '20	$C_L = 15$ pF, $R_L = 400$ Ω		12	22		8	15
'30			13	22		8	15
'H00			5.9	10		6.2	10
'H04			6	10		6.5	10
'H10	$C_L = 25$ pF, $R_L = 280$ Ω		5.9	10		6.3	10
'H20			6	10		7	10
'H30			6.8	10		8.9	12
'LS00, 'LS04 'LS10, 'LS20	$C_L = 15$ pF, $R_L = 2$ kΩ		9	15		10	15
'LS30			8	15		13	20
'S00, 'S04	$C_L = 15$ pF, $R_L = 280$ Ω		3	4.5		3	5
'S10, 'S20	$C_L = 50$ pF, $R_L = 280$ Ω		4.5			5	
'S30, 'S133	$C_L = 15$ pF, $R_L = 280$ Ω		4	6		4.5	7
	$C_L = 50$ pF, $R_L = 280$ Ω		5.5			6.5	

As an example of this, the section of a data sheet given in Table 6.1 quotes propagation delays in nanoseconds for a range of gates, assuming a supply voltage of 5 V and an ambient temperature of 25° C. The propagation delay for each gate in a 7404 hex inverter (which is 14-pin dual-in-line integrated circuit containing six NOT gates) can be seen to be typically 12 ns for a low to high output and 8 ns for high to low output.

6.4 Diode–transistor logic (DTL)

It is possible to construct an AND gate very simply by using diodes and resistors as shown in Fig. 6.5.

This shows a three-input AND gate with inputs A, B and C. Assuming that a logic 0 level is 0 V and a logic 1 level is +5 V (at the inputs), the possible outputs can be calculated. If all of the inputs A, B, C are at 5 V, then there is no forward bias and no current will flow. As a result, the output voltage will be equal to the supply voltage of 5 V. This situation is shown in Fig. 6.6.

If any one of the inputs is at logic 0 then current will flow from the supply, through the resistor R_1 and the forward biased diode, producing a voltage drop across the diode which is approximately 0.7 V. This is shown in Fig. 6.7.

Input C has been reduced to logic 0 but the output voltage corresponding to a logic 0 has increased to 0.7 V. If this logic 0 output is fed to another identical gate, then the output logic 0 from the second gate would be a voltage of 1.4 V. Clearly this should not be allowed

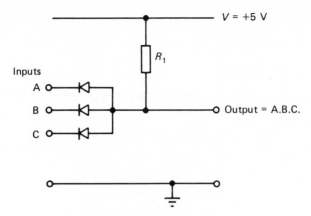

Fig 6.5   Diode-resistor logie AND gate

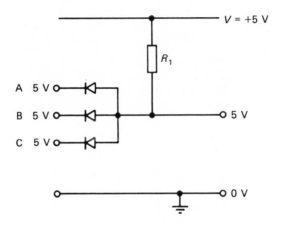

Fig 6.6   AND gate with all inputs at logic 1

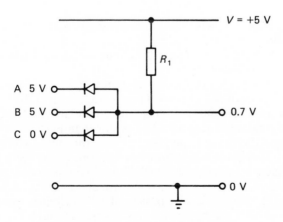

Fig 6.7   AND gate with one input at logic 0

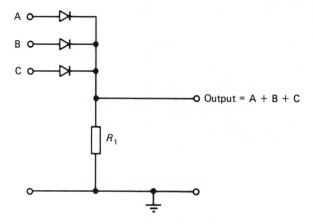

Fig 6.8   Diode-resistor logic OR gate

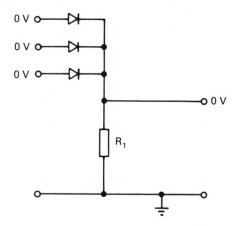

Fig 6.9   OR gate with all inputs at logic 0

to happen as recognising a 0 will become increasingly difficult. An OR gate can be constructed in a similar way, as shown in Fig. 6.8.

If all of the inputs A, B, C are at logic 0 (0 V), then the output logic 0 will be at 0 V also. This condition is shown in Fig. 6.9.

If any of the inputs is at logic 1 (5 V), then current will flow into the circuit at that input through the diode and resistor R_1. There will be a voltage drop of 0.7 V across the forward biased diode, so that the actual output voltage corresponding to a logic 1 will be degraded to 4.3 V from the original input 1 of 5 V. This is shown in Fig. 6.10.

If this degraded logic 1 were fed to an identical gate, the output voltage from the second gate would be 3.6 V. This cannot be allowed to happen, as the logic 1 level would soon become indistinguishable from a logic 0. The solution to this is to add a buffer to make sure that the output logic levels are consistent, irrespective of how many gates are cascaded. The inverting buffer or NOT gate previously considered will fulfil this function. This forms the basis of a diode–transistor logic gate which is illustrated in Fig. 6.11.

The AND gate previously considered is followed by a NOT gate which produces a

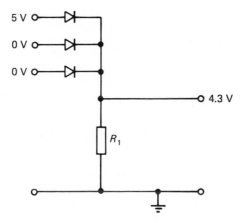

Fig 6.10 OR gate with one input at logic 1

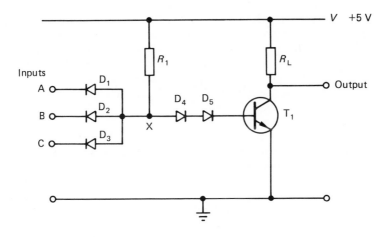

Fig 6.11 Diode transistor logic NAND gate

NAND gate overall. The diodes D_4 and D_5 are called **level shifting diodes** and their function will be considered shortly. If all inputs to the NAND gate are high (logic 1) then transistor T_1 will be forward biased and base current will flow from the supply via R_1, D_4 and D_5, switching T_1 *on* and producing a low (logic 0) output. The output voltage will be $V_{CE(sat)}$, as previously explained for the NOT gate. The voltage at point X is determined by the number of forward biased junctions between X and ground, which in this case is 3. If a drop of 0.7 V per junction is assumed, then $V_X = 2.1$ V. If any of the inputs to the gate are low (logic 0) then the voltage at point X will be given by:

$$V_X = V_{in(low)} + 0.7$$

It has already been shown that V_X must be 2.1 V to switch the transistor *on*. $V_{in(low)}$ therefore has a maximum value of 1.4 V to prevent this from happening. If the gate is fed from the output of a similar gate, this voltage will be in the order of 0.3 V ($V_{CE(sat)}$), giving:

$$V_X = 0.3 + 0.7 = 1.0 \text{ V}$$

which is 1.1 V less than the 2.1 V required to give a false output. If the level shifting diodes D_4 and D_5 were not present, then the voltage required at X to forward bias T_1 would only be 0.7 V, and T_1 would therefore be permanently *on* regardless of the input. With the diodes present, a margin is introduced which protects the gate against false outputs due to electrical noise spikes which can increase the input voltage $V_{in(low)}$ to a value that causes T_1 to switch *on*. This margin is a measure of the **noise immunity** of the stage, and is one of the design factors that would be specified on a data sheet for a particular gate. In the situation where inputs A, B and C are all high, say 5 V, then T_1 is saturated and V_X equals 2.1 V. If a noise voltage on any input, say A, were to reduce the 5 V, input level to less than 1.4 V, than V_X would reduce below 2.1 V and the transistor would begin to switch *off*. This gives a noise margin of 5 V − 1.4 V = 3.6 V between the normal high voltage (logic 1) and a noisy input, before a false output occurs. As this margin is greater than the margin calculated for a positive noise spike, then the smallest margin should be the value used if the gate is to be implemented in a noisy environment. Note, however, that 5 V is not a typical logic 1 voltage level for a bipolar device. The logic 1 level will be typically 3.4 V, which would mean that the noise margin would (for negative noise spikes) be 3.4 V − 1.4 V, = 2.0 V. It is worth noting that increasing the number of diodes improves the noise immunity for positive noise voltages, but reduces the noise margin for negative voltage spikes. For example, if an extra diode were to be included in the original case, the noise margin (low) would increase from 1.1 V to 1.8 V and the noise margin (high) would reduce from 3.8 V to 2.9 V. A compromise must be reached, and normally 2 or 3 diodes are used. In the event of a gate having to be used in an environment where very large noise levels exist, for example as part of a motor speed control circuit, there is a logic family called **high threshold logic** which uses a 15 V supply instead of 5 V. The higher supply voltage permits a larger number of noise immunity (level shifting) diodes to be used, and increases both high and low noise margins when compared to the circuit previously considered.

To determine how many similar gate inputs can be driven from the DTL gate, some calculations are necessary. The maximum number of gate inputs that can be connected to the output of a gate is called the **fan-out** of the gate. The circuit of Fig. 6.11 is redrawn in Fig. 6.12 with component values added.

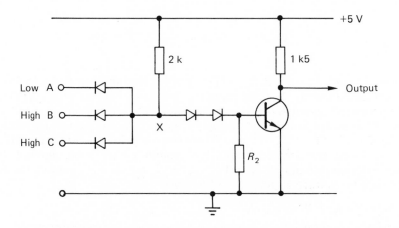

Fig 6.12   DTL NAND gate for fan-out calculation

If it is assumed that all of the successive gate inputs connected to the gate output are *low*, i.e. at logic 0, then the worst possibility can be catered for. A high input will not draw current. This is illustrated in Fig. 6.13.

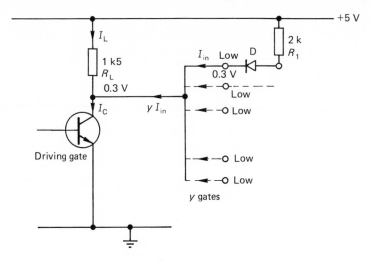

Fig 6.13   Calculation of fan-out

If $V_{in(low)} = 0.3\,V$
then $V_X = V_{in(low)} + V_D = 1.0\,V$
and $I_{in} = (V_{CC} - V_X)/R_1 = (5 - 1)/2\,k\Omega = 2\,mA$
Also $I_C = I_L + y\,I_{in}$
But $I_L = (V_{CC} - V_{in(low)})/R_L = (5 - 0.3)/1.5\,k\Omega = 4.7/1.5\,mA = 3.13\,mA$
so $I_C = (3.13 + 2y)\,mA$

In addition,

$$I_C = h_{FE}I_B$$

where h_{FE} is the current gain of the transistor.
Assuming a minimum value of h_{FE} to give the worst case condition:

$$h_{FE(min)} = 30$$

then

$$I_B = I_c/h_{FE} = (3.13 + 2y)/30\,mA$$
$$I_B = (0.104 + 0.067y)\,mA$$

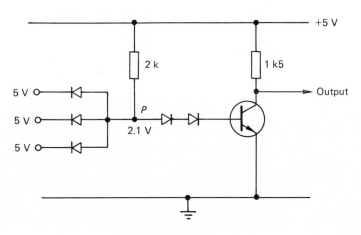

Fig 6.14   DTL NAND gate

If I_B is calculated from the circuit values, it is necessary to consider the supplying gate in the *on* condition. For this, all inputs are high and a voltage of 2.1 V is on point P as shown in Fig. 6.14 being ($2 V_D + V_{BE(sat)}$) which is approximated to 3 $V_{BE(sat)}$.

The current through the 2kΩ resistor is equal to the base current as all three inputs are reverse biased. Thus

$$I_B = \frac{V_{CC} - 3V_{BE(sat)}}{R_1}$$

$$I_B = (5 - 2.1)/2k\Omega = 2.9/2\,mA = 1.45\,mA$$

so $I_B = 1.45 = 0.104 + 0.067y$

where *y* is the number of gates,

$$y = (1.45 - 0.104)/0.067 = 1.346/0.067 = 20.09$$

so that with the circuit of Fig. 6.11 the fan-out would be 20.

The circuit can be connected to the inputs of twenty similar gates with the component values used.

6.5 Transistor-transistor logic (TTL)

The circuit of a transistor–transistor logic (TTL) NAND gate is given in Fig. 6.15.

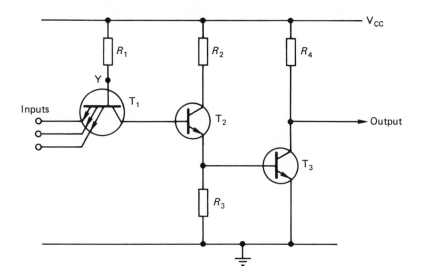

Fig 6.15   TTL NAND gate

The operation of a TTL NAND gate is similar in many ways to that of the DTL NAND gate previously considered, and it is convenient to try to understand how the TTL gate works in terms of modified DTL. The first difference is at the input, where the three input diodes in the DTL circuit are replaced by a multi-emitter transistor (MET) with three emitters acting as inputs. If any input is taken low, then the base–emitter junction on transistor T_1 will become forward biased and T_1 will saturate. However, the magnitude of the base voltage will be fairly low, as the base current through R_1 will reduce the potential of point Y to ($V_{in} + V_{BE(sat)}$). If $V_{in(low)}$ is assumed to be 0.3 V as before, then $V_Y = 0.3 + 0.7 = 1.0$ V. This is not high enough to allow T_2 and T_3 to switch *on*. If T_2 and

T_3 were *on*, the potential at the base of T_2 would be 1.4 V. As T_1 is assumed to be in saturation, the collector voltage on T_1 will be given by:

$$V_C = V_{CE(sat)} + V_{in(low)}$$
$$= 0.3 + 0.3 = 0.6 \text{ V}$$

Clearly, as this is less than the required value of 1.4V, T_2 and T_3 must be *off*, giving a high output from the NAND gate. If all inputs are high, as shown in Fig. 6.16, then the base–emitter junction of T_1 is reverse biased and no current flows in the emitter.

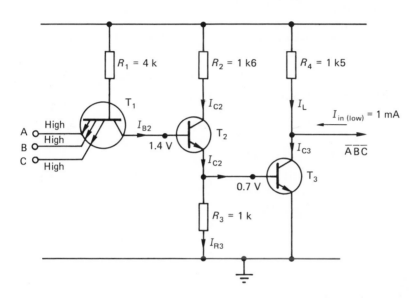

Fig 6.16 NAND gate operation

The base–collector junction, however, will be forward biased, and acts as if it were an ordinary diode D_4 in the DTL circuit of Fig. 6.11. Current is therefore supplied to the base of transistor T_2, switching it *on*. Emitter current flows in resistor R_3 and the base voltage rises on T_3, switching it *on*. The output voltage from T_3 will be equal to $V_{CE(sat)}$, or approximately 0.3 V. The base–emitter junction of transistor T_2 effectively replaces diode D_5 in the DTL circuit of Fig. 6.11. Thus the gate shown in Fig. 6.15 acts as a NAND gate as required. At first glance it would seem that it only reproduces the DTL gate considered but in a different form. There are, however, distinct advantages in employing a TTL gate in preference to DTL. The advantages are:

(1) Higher speed of operation.
(2) Increased fanout.

An important point to note is that transistor T_1 in Fig. 6.16 will only saturate if an input is taken low. If a logic 1 is applied to all emitters of T_1 *or* if all inputs are disconnected, T_1 will remain *off*. Thus an open or disconnected input acts as if it were a logic 1. An open input is said to **float high**. Unused inputs should not be left floating, as they will be susceptible to stray noise. They should either be strapped to V_{CC} (NAND/AND) or GND (OR/NOR). Alternatively, unused inputs can be connected together. If an unused input is connected to V_{CC}, it is recommended that this should be via a 1kΩ resistor to limit any current spikes due to supply variations.

In fact the TTL NAND gate depicted in Fig. 6.16 is not the final design – there are improvements that can be added to upgrade its performance even more. As it stands, however, the speed is considerably increased because of the gain provided by the transistors T_1 and T_2. If it is assumed that inputs A, B and C are all high and then that A is suddenly reduced to a low logic level, T_1 becomes saturated. The collector current of T_1 drains away base charge from T_2 at a much greater rate than would be the case with a passive diode, because the gain on T_1 provides a large value of collector current. As a result of this, T_2 and thus T_3 switch *off* rapidly. This reduction in propagation delay increases the maximum frequency at which the gate may be used.

The fan-out of the TTL gate can be determined by a similar method to the DTL fan-out calculation. Component values must be allocated and are indicated in Fig. 6.16.

The gate output is low because A, B and C are all high. If any input were low then, assuming $V_{in} = 0.3\,\text{V}$, I_{in} would be given by:

$$I_{in} = (V_{CC} - V_{BE(sat)} - V_{in})/R_1 = (5 - 0.7 - 0.3)/4\,\text{k}\Omega = 4/4\,\text{mA} = 1\,\text{mA}$$

The gate with low output shown in Fig. 6.16 would give:

$$I_{B_2} = (V_{CC} - 3\,V_{BE(sat)})/R_1 = (5 - 2.1)/4\,\text{k}\Omega = 23.9/4\,\text{mA} = 0.725\,\text{mA}$$

because the base–collector junction of T_1 acts as a diode, so that $I_{B_1} = I_{C_1} = I_{B_2}$.

The collector current of T_2 is given by:

$$I_{C_2} = (V_{CC} - V_{CE_2(sat)} - V_{BE_3(sat)})/R_2 = (5 - 0.3 - 0.7)/1.6\,\text{k}\Omega = 4/1.6\,\text{k}\Omega = 2.5\,\text{mA}$$

because T_2 is saturated.

The current through R_3 is given by:

$$I_{R_3} = V_{BE_3}/R_3 = 0.7/1\,\text{k}\Omega = 0.7\,\text{mA}$$

The base current of T_3 is given by:

$$\begin{aligned} I_{B_3} &= I_{E_2} - I_{R_3} = I_{B_2} + I_{C_2} - I_{R_3} \\ &= 0.725 + 2.5 - 0.7 \\ &= 2.525\,\text{mA} \end{aligned}$$

For T_2 to be fully on,

$$h_{FE}I_{B_2} > I_{C_2}$$

Assuming $h_{FE(min)} = 30$ for T_2,

$$30 \times 0.725 = 21.75\,\text{mA} \gg 2.5\,\text{mA}$$

Assuming $h_{FE(min)} = 30$ for T_3,

$$\begin{aligned} h_{FE}I_{B_3} &= 30 \times 2.525\,\text{mA} \\ &= 75.75\,\text{mA} \end{aligned}$$

which will cause T_3 to saturate.

The current I_L through R_4 is given by

$$I_L = \frac{V_{CC} - V_{CE(sat)}}{R_4}$$

$$I_L = \frac{5 - 0.3}{1.5}\,\text{mA}$$

$$= \frac{4.7}{1.5}\,\text{mA}$$

$$= 3.13\,\text{mA}$$

The collector current of T_3 is given by

$$I_{C_3} = I_L + x$$

where x represents the fan-out, and remembering that $I_{in} = 1\,mA$, so

$$x = 75.75 - 3.13 = 72.62$$

For the circuit of Fig. 6.16, the fan-out would therefore be 72.

This is a great improvement on a direct DTL equivalent, and is a major advantage of TTL. However, a fan-out of 72 would require that the collector current of the final stage is in excess of 75 mA which is high. Other considerations would probably reduce this value.

6.6 Totem pole output

The TTL circuit shown in Fig. 6.15 can be improved by modifying the output stage. When a logic 0 output changes to a logic 1 output in this circuit, transistor T_3 switches *off* and the collector voltage of T_3 rises. The rate at which this voltage increases is determined by the output capacitance C_{out} and the load resistance R_4. This is illustrated in Fig. 6.17.

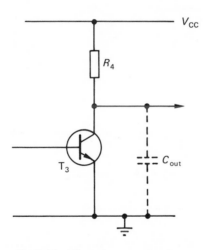

Fig 6.17 Effect of output capacitance

If T_3 switches *off* (i.e. an open circuit between collector and emitter), then V_{out} will rise at a rate determined by the time constant $T = C_{out} R_4$.

Ideally, if R_4 could be reduced to zero when T_3 switches *off*, this would reduce the time constant to zero and eliminate the delay, giving a significant increase in speed. By replacing R_4 with a transistor and small resistor, an **active** load for transistor T_3 is introduced. The newly introduced transistor can be switched on when T_3 switches off, greatly reducing the time constant and increasing the switch off speed. The output stage with this arrangement is known as a **totem pole output** and is standard in most TTL circuits. Figure 6.18 illustrates a NAND gate using this design.

The **passive** resistor in the collector of the output transistor has been replaced by an **active** load consisting of transistor T_4 and low value resistor R_4. Transistor T_2 acts as a phase-splitter, so that when T_2 is *on*, T_3 is *on* and T_4 is *off*; and when T_2 is *off*, T_3 is *off* and T_4 is *on*. The fact that T_4 is *off* when T_3 is *on* effectively gives a high impedance load to T_3, reducing power dissipation. When T_3 switches *off*, the action of T_4 switching *on* reduces the effective load resistance of T_3 and reduces the switch-off time as previously explained. The resistance R_4 is included to limit the current during the brief period that both transistors T_3 and T_4 are conducting, because in fact T_4 will switch *on* before T_3 can switch *off* when the output voltage is going from low to high. Unfortunately, during this brief period, current is limited only by the low resistance of R_4, and so a large short duration current pulse is extracted from the supply. This would not in itself be a problem if it were not for the inductance of the supply lines. The combination of this inductance and the large short pulse can cause a corresponding voltage pulse, reducing the supply voltage. When a number of TTL gates are connected to the same supply, the problem can become serious and the

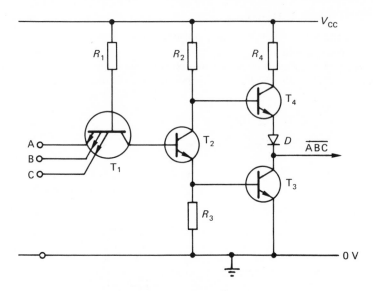

Fig 6.18   Totem-pole output

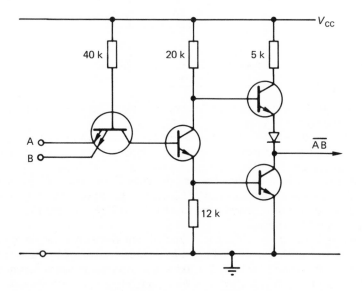

Fig 6.19   High value resistors – low power consumption – low speed

standard solution is to connect a disk ceramic capacitor to each separate TTL integrated circuit on the printed circuit board.

To return to the totem pole output, it should be noted that R_4 also provides a degree of short circuit protection. The diode D is included to ensure that when T_3 is conducting, T_4 is *off*. If it were not present, the combination of $V_{CE(sat)}$ for T_2 and $V_{BE(sat)}$ for T_3 would cause T_4 to switch on.

To summarise, the addition of an active load allows higher operating speeds without increasing power dissipation. The resistor values used in the circuit of Fig. 6.18 determine both the speed and power dissipation of the gate. TTL gates are categorised by these factors, and Figs. 6.19, 6.20, and 6.21 illustrate three of these categories.

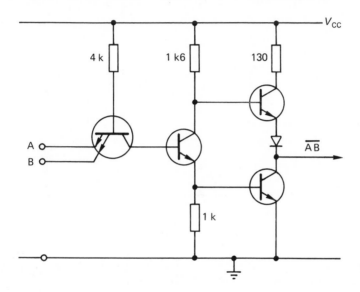

Fig 6.20   Standard TTL – lower resistor values – higher speed

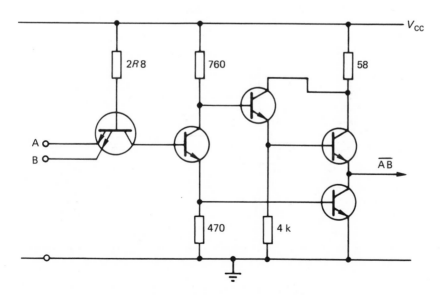

Fig 6.21   High speed – high power consumption – low resistor values

The circuit shown in Fig. 6.21 includes an additional stage which permits even higher operating speeds.

6.7   Other TTL gates

A TTL NOR gate is illustrated in Figure 6.22.

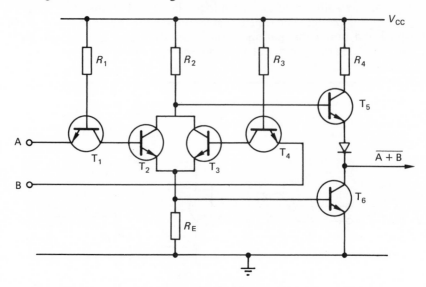

Fig 6.22   TTL NOR gate

If either input A or B is at logic 1, then the output will be at logic 0.
Figure 6.23 illustrates an AND-OR-INVERT (AOI) gate.

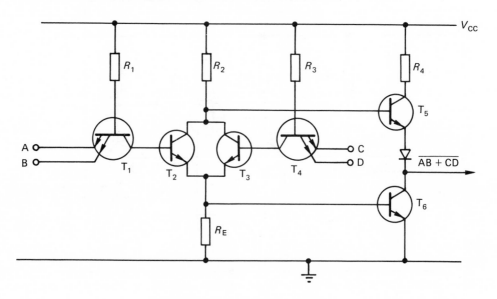

Fig 6.23   TTL AND – OR – INVERT gate

The equivalent in terms of two AND gates and a NOR gate is given in Fig. 6.24.

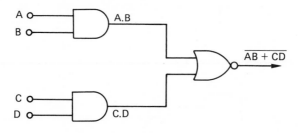

Fig 6.24   AOI equivalent circuit

An open-collector NAND gate is shown in Fig. 6.25.

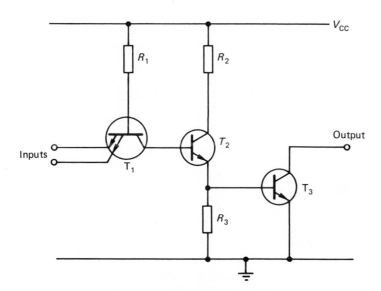

Fig 6.25   Open collector NAND gate

The components in the collector of the standard TTL gate are omitted, to form an open-collector gate which has the NAND function but which requires an external load resistance to operate. This resistance is usually referred to as a **pull-up resistor**. Open-collector gates are useful where gate outputs are wired together to produce the effect of an AND function at the junction of the outputs, without the need for an actual AND gate.

   Tristate gates are designed so that instead of 0 and 1 being the only possible output states, a third 'high impedance' state is introduced. The operation of a tristate can be illustrated by a simple modification to a TTL NAND gate as shown in Fig. 6.26.

   If the control input is maintained at a logic 1 level, then the gate will operate as a conventional TTL inverter. If the control input is set to zero, then all drive current is removed from the output transistors and they are both switched *off*. The effect of this as far as the output circuit is concerned is shown in Fig. 6.27.

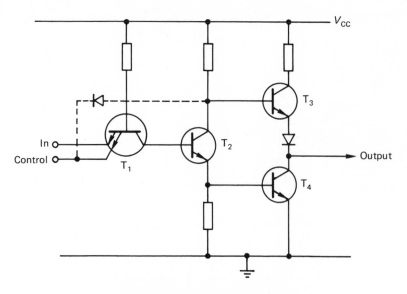

Fig 6.26   Operation of a tristate gate

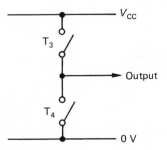

Fig 6.27   Equivalent circuit of output stage of a tristate

T_3 and T_4 are represented as open switches to correspond to their *off* condition. This means that the output is floating, or in a high impedance state. This provides isolation to any successive devices connected to the output of the gate. The symbol for an inverting tristate is given in Fig. 6.28, whereas Fig. 6.29 shows a non-inverting tristate.

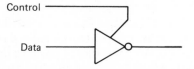

Fig 6.28   Inverting tristate

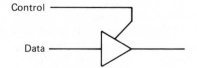

Fig 6.29   Non-inverting tristate

The tristate is widely used in microprocessor systems to permit common bus lines to be used, as shown in Fig. 6.30.

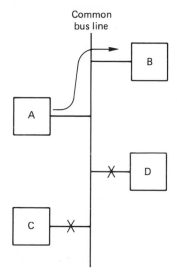

Fig 6.30   Tristate operation

If information is to flow from A to B, then the connections to C and D from the common busline must be open circuited, i.e. in the high impedance state. Tristates in the positions marked with a cross could be used to do this.

If data flow is required in two directions, then a bidirectional tristate must be used, as shown in Fig. 6.31.

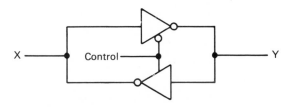

Fig 6.31   Bidirectional tristate

The common configuration shown allows data to flow from X to Y if the control input is at logic 0, and from Y to X if the control input is at logic 1. The small circle or bubble on the upper control input represents an inverted input, or an 'active low' input, which means that the tristate is enabled by a low (logic 0) input value.

6.8 Schottky TTL

TTL is a form of saturating logic. The resulting limitation on speed has already been explained. The problem of switching off a saturated transistor can be avoided by preventing the transistor from going into deep saturation in the first place. A device which does not suffer from charge storage problems is the **Schottky diode**. This is *not* a semiconductor p–n junction diode, and has a smaller forward voltage drop than normal. It can be connected between the base and collector of an ordinary bipolar transistor as shown in Fig. 6.32.

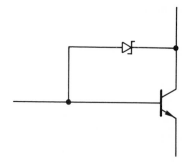

Fig 6.32   Schottky diode modification to a bipolar transistor

When the transistor switches *on*, the voltage on the collector falls and at some point the diode will start to conduct, diverting current away from the base and limiting the depth of saturation. The lower forward voltage drop of the Schottky diode means that it will conduct before an equivalent bipolar diode. The lack of charge storage in the Schottky diode means that there is no difficulty in preventing it from conducting when the transistor is switched *off*. As a result the Schottky modified transistor, whose symbol is given in Fig. 6.33, will switch off much quicker than the transistor alone, reducing the propagation delay if it is used in a gate circuit and increasing the switching speed. Schottky TTL devices can operate up to 100 MHz.

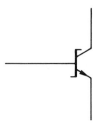

Fig 6.33   Schottky transistor

A Schottky TTL NAND gate is illustrated in Fig. 6.34.

It is possible to combine low power TTL and Schottky TTL circuits to give low power Schottky TTL. This has the advantage of low power dissipation and high speed operation. Many integrated circuit distributors now concentrate exclusively on low power Schottky devices when supplying TTL. Different types of TTL are distinguished as follows:

(i)   74XX implies standard TTL, e.g. 7400
(ii)   74HXX implies high speed TTL, e.g. 74H04
(iii) 74LXX refers to low power TTL, e.g. 74L08
(iv)   74SXX implies Schottky TTL, e.g. 74S10
(v)   74LSXX refers to low power schottky TTL, e.g. 74LS76

6.9   PMOS and NMOS logic families

Metal oxide semiconductor (MOS) logic families are based on two main transistor types, PMOS and NMOS. PMOS transistors are p-channel devices and have the symbol given in Fig. 6.35(a). NMOS transistors are n-channel devices and have the symbol given in Fig. 6.35(b).

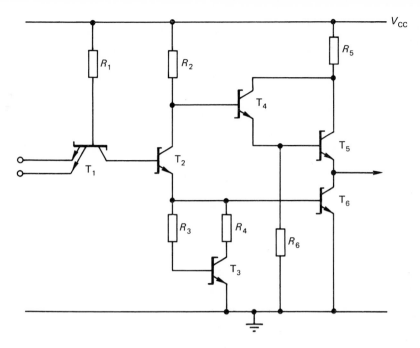

Fig 6.34   Schottky TTL NAND gate

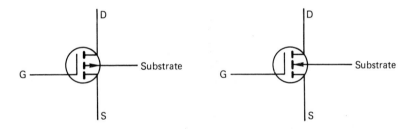

Fig 6.35   (a) PMOS transistor symbol; (b) NMOS transistor symbol

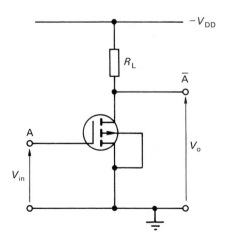

Fig 6.36   PMOS NOT gate (passive load)

It is possible to produce a NOT gate using a p-channel transistor as shown in Fig. 6.36. In integrated circuit form, the load resistance R_L is usually more difficult to produce than a transistor, so IC implementation usually has an active load as shown in Fig. 6.37.

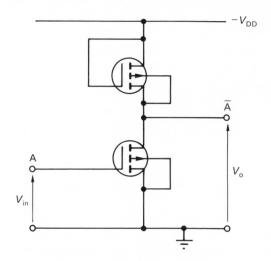

Fig 6.37   PMOS NOT gate (active load)

The circuit of a PMOS NOR gate is given in Fig. 6.38.

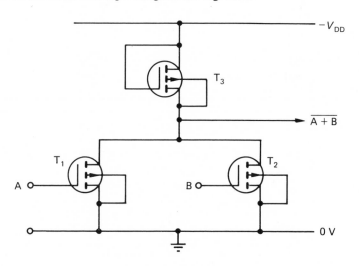

Fig 6.38   PMOS NOR gate

The transistor T_3 will act as an active load as in the NOT gate previously described. If either input A or B is negative, then T_3 will conduct and the output voltage will be close to 0V. If both A and B are at 0V, then T_1 and T_2 will be switched *off* and the output voltage will be close to $-V_{DD}$. If this is interpreted with 0V = logic 0 and $-V_{DD}$ = logic 1 (negative logic), then the gate provides the NOR function.

The circuit for a PMOS NAND gate is given in Fig. 6.39.

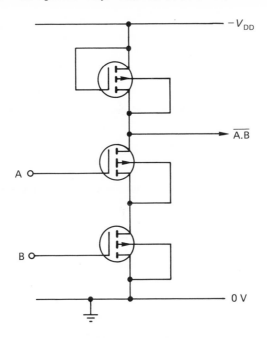

Fig 6.39   PMOS NAND gate

PMOS gates are enhancement-mode devices which are fairly easy to manufacture at high density and low cost with limited speed.

NMOS gates have circuits that are very similar to those described for PMOS devices with reversed polarity supplies. These are given in Figure 6.40. The advantage of NMOS is that it will operate at higher frequencies and dissipate less power than PMOS although this is at the expense of higher production costs. NMOS devices predominate because of these advantages.

6.10 Complementary metal oxide semiconductor (CMOS) logic

Complementary metal oxide semiconductor logic, or CMOS logic, utilises both p and n-channel MOS field effect transistors (MOSFETS), which has the effect of producing logic gates with very low power consumption and high noise immunity. A two-input NAND gate using positive logic is illustrated in Figure 6.41.

Note that T_1 and T_2 are p-channel devices and that T_3 and T_4 are n-channel. A logic 0 input on either A or B will cause one or both of the p-channel transistors to switch on, turning off one or both of the n-channel transistors and giving an output voltage approximately equal to V_{DD} (logic 1). Only if both A and B inputs are high will both T_1 and T_2 turn *off* and T_3 and T_4 turn *on*, reducing the output to a low logic 0 level. A CMOS NOR gate is given in Fig. 6.42.

If either A or B input is at logic 1 then either one or both of T_3 and T_4 will be *on*. The output will therefore be low. If both A and B are at logic 0, then both T_1 and T_2 will be *on* and T_3 and T_4 *off*, giving a high output level.

The major problem with CMOS logic is the restriction on speed imposed by the MOSFET which are much slower than their bipolar counterpart. Also, if a higher frequency operation is attempted with CMOS gates, the power dissipation increases

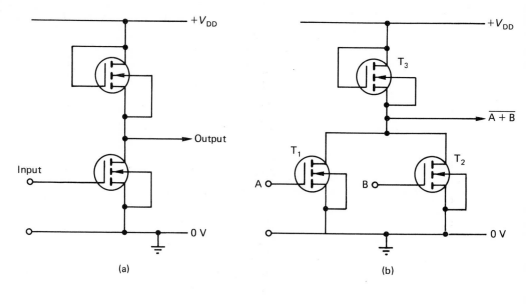

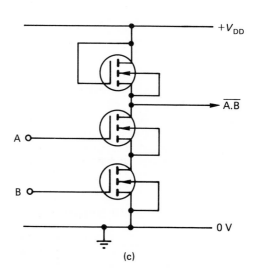

Fig 6.40 (a) NMOS gate (active load); (b) NMOS NOR gate; (c) NMOS NAND gate

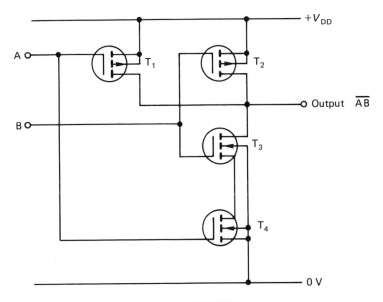

Fig 6.41 CMOS NAND gate

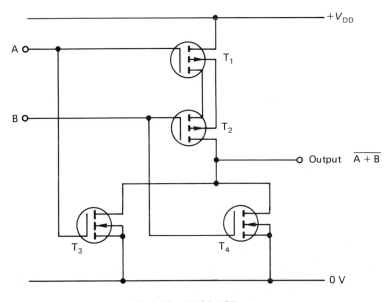

Fig 6.42 CMOS NOR gate

considerably. The reason for this is that every time a low to high transition occurs at the output of a CMOS gate, the output capacitance must be charged by a current pulse from the supply. The magnitude of the output capacitance will be determined by the number of CMOS inputs being supplied. As the frequency increases, more current pulses per second are drawn from the supply and the power consumption and dissipation increases. CMOS devices have greater noise immunity than TTL, however, and will operate over a wide range of supply voltages. Like all MOS devices, CMOS can be damaged by static electrical charges. This is because there is a very thin layer of insulation between the gate of the transistor and the p- or n-channel which can easily be punctured by a high voltage. CMOS gates usually have input protection circuits to reduce the risk of damage in this way. However, more care in handling is required than with bipolar devices. An advantage of CMOS is that it is less temperature sensitive than bipolar technologies. The packing density of CMOS logic is less than with either PMOS and NMOS. CMOS microprocessors are rare as a result of this. A propagation delay of 40 ns is typical in a CMOS gate. Noise immunity is high, usually in the region of 45% of the supply voltage. CMOS finds wide application in industrial environments, motor vehicles and portable equipment including watches and space vehicles, where low power consumption is essential.

6.11 High speed CMOS (HCMOS)

The HCMOS logic family is a relatively recent development which will probably be used extensively in future years. It combines many of the features of CMOS and TTL. The basic difference between CMOS and HCMOS is that an advanced silicon gate technology is used. This reduces parasitic capacitance and allows higher speed of operation. Together with its increased transistor gain (which allows faster discharge of parasitic capacitance), HCMOS speeds are compatible with low power Schottky TTL. This is achieved without loss of the CMOS feature of low power dissipation. The HCMOS family uses the same pin-outs as TTL, and in most cases are exact equivalents and are fully compatible with either TTL or CMOS gates. A typical value of quiescent power dissipation per gate is $10 \mu W$, which is considerably less than that of the equivalent LSTTL gate.

An example of HCMOS is the 74HC08 quad two-input AND gate illustrated in Figure 6.43.

The propagation delay is 9.5 ns, and it has a fan-out of ten LSTTL loads. The power consumption is $10 \mu W$ maximum at room temperature with a 5 V supply voltage.

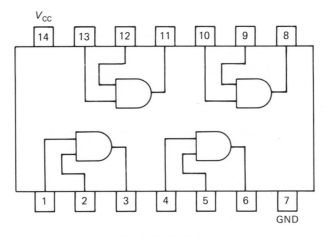

Fig 6.43   HCMOS AND gate

6.12 Emitter-coupled logic (ECL)

Emitter-coupled logic (ECL) gates are designed to be non-saturating, avoiding delays associated with bipolar transistor storage time as experienced by TTL, and thus operate at a much higher frequency. An ECL OR/NOR gate is shown in Fig. 6.44.

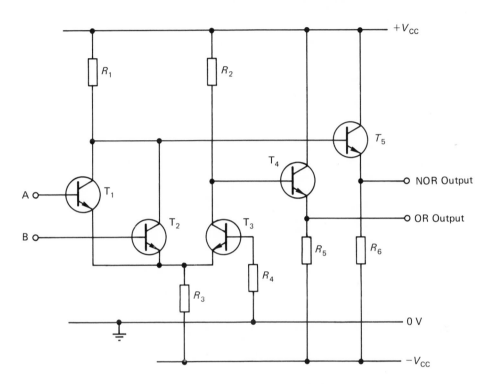

Fig 6.44   ECL OR/NOR gate

Transistors T_1 and T_2 are connected in parallel with common collector resistor R_1. If inputs A and B are low, transistors T_1 and T_2 switch *off* producing a high potential on their common collector. This high potential will cause T_5 to conduct, giving a logic 1 on the NOR output as expected. The emitter resistor R_3 is common to T_1, T_2, and T_3, and is of sufficiently high value to act as a constant current source. While T_1 and T_2 are *off*, current flows from V_{CC} via R_2 to T_3 which switches off T_4 giving a logic 0 on the NOR output. Transistor T_3 will switch *off*, T_4 will switch *on* and a logic 1 will appear on the OR output as expected. The fact that the power supply always maintains a constant current input which is merely switched from one arm of a parallel circuit to the other means that internal noise due to large switching transients is considerably reduced with ECL.

6.13 Integrated injection logic (I²L)

Integrated injection logic (I²L) is a bipolar logic family designed to provide high component density and high speed operation.
An I²L NAND gate is shown in Fig. 6.45.

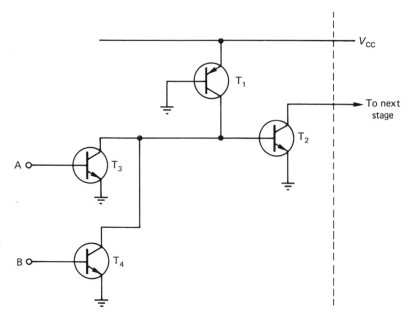

Fig 6.45   I²L NAND gate

Transistor T_1 operates as a current source. If inputs A and B are both low (logic 0), then transistors T_3 and T_4 are cut off and current flows to the base of T_2, switching it *on*. The gate output is therefore low. If either A or B are high, than T_3 or T_4 saturates and the current is removed from the T_2 base, switching off T_2 and giving a high gate output, i.e. the circuit operates as a NAND gate. A major advantage of I²L is that it does not require any resistors. In conjunction with the fact that it is possible to manufacture the transistors with common regions (eg. T_1 collector and T_2 base), the fabrication process is very much simplified.

6.14 Understanding data sheets

In the discussion which follows superscripts in the form of a number enclosed in square brackets refer to the numbers annotated on Table 6.2.

Data for the 7400 [1] NAND gate is given in Table 6.2. This will be used as an example of a typical data sheet, and some guidance as to how to interpret its contents will be given.

Supply voltage V_{CC}
The nominal supply voltage (NOM) [2] is given as 5 V. It is permissible to drop 0.25 V to 4.75 V (MIN) [3] or increase to 5.25 V (MAX) [4] while still maintaining correct operation of the gates.

Operating Temperature
The typical operating temperature is quoted as 25°C [5] with a lower limit of 0°C [6] and an upper limit of 70°C [7]. Note that there is an equivalent device, the 5400 NAND [8], which is identical logically but has an operating temperature of − 55°C to + 125°C.

Input high voltage (V_{IH})

The value of 2 V [9] quoted is the minimum voltage that will be accepted by the gate as a logic 1 level at any input.

Input low voltage (V_{IL})

The value of 0.8 V [10] quoted is the maximum voltage that will be accepted as a logic 0 level at any input.

Output high voltage (V_{OH})

The typical voltage for a logic 1 output is 3.4 V [11]. The minimum voltage will be 2.4 V [12].

Output low voltage (V_{OL})

The typical voltage for a logic 0 output is 0.2 V [13]. The maximum voltage will be 0.4 V [14].

Input High Current (I_{IH})

When a logic 1 input is applied at a specified voltage, the quoted current will flow. With $V_{in} = 2.4$ V the input current will be 40 μA [5].

Input Low Current (I_{IL})

When a logic 0 input is applied at a specified voltage the quoted current will flow. With $V_{in} = 0.4$ V the input current will be $-$ 1.6 mA [16]. The significance of the negative sign is that the current flows *out* of the input, in other words positive current flow is *into* the gate.

Output High Current (I_{OH})

This is the current that flows out of the gate when the output is at logic 1. In this case $I_{OH} = -$ 400 μA [17]. The negative sign implies an outward flow of current.

Output Low Current (I_{OL})

This is the current that flows at the gate output when the output is at logic 0. The value in this case is 16 mA [18].

It is useful to represent the input and output voltage levels derived from the data sheet in diagrammatic form. This is given in Fig. 6.46.

The range of permitted supply voltages is given in Fig. 6.46 (a). Figure 6.46 (b) gives input and output voltage levels. Any voltage greater than $V_{OH(min)}$ can be regarded as a valid logic 1 output. Any voltage less than $V_{OL(max)}$ can be regarded as a valid logic 0 output. The region between these is undefined. Any voltage greater than V_{IH} will be accepted as a valid logic 1 input and any voltage less than V_{IL} will be regarded as a valid logic 0 input. The region between V_{IH} and V_{IL} is undefined. If the minimum possible output voltage corresponding to a logic 1 is $V_{OH(min)}$ and the minimum possible logic 1 input voltage is V_{IH}, then it is possible to have up to ($V_{OH(min)} - V_{(IH)} = 2.4 - 2 = 0.4$ V of negative-going noise before the undefined area is entered. This is called the **noise margin (high)** and is illustrated in Fig. 6.47.

If the maximum possible output voltage corresponding to a logic 0 is $V_{OL(max)}$, and the maximum possible logic 0 input voltage is V_{IL}, then it is permissible to have up to ($V_{IL} - V_{OL(max)} = 0.8 - 0.4 = 0.4$ V of positive going noise before the undefined area is entered. This is called the **noise margin (low)** and is illustrated in Fig. 6.48.

The fan-out of the gate can be calculated by dividing I_{OL} by I_{IL}, which in this case gives 16 mA/1.6 mA $= 10$. This is the fan-out for the *low* condition.

The fan-out of the gate in the *high* output condition is determined by dividing I_{OH} by I_{IH}, which gives 400 μA/40 μA $= 10$.

Table 6.2   Typical data sheet. (Courtesy of Texas Instruments Inc.)

Recommended operating conditions

			54 FAMILY 74 FAMILY	SERIES 54 SERIES 74	
				'00, '04, '10, '20, '30	
			MIN	NOM	MAX
Supply voltage, V_{CC}		54 Family	4.5	5	5.5
		74 Family	4.75	5	5.25
High-level output current, I_{OH}		54 Family			−400
		74 Family			−400
Low-level output current, I_{OL}		54 Family			16
		74 Family			16
Operating free-air temperature, T_A		54 Family	−55		125
		74 Family	0		70

Electrical characteristics over recommended operating free-air temperature range (unless otherwise noted)

				SERIES 54 SERIES 74		
				'00, '04, '10, '20, '30		
PARAMETER		TEST FIGURE	TEST CONDITIONS†	MIN	TYP‡	MAX
V_{IH}	High-level input voltage	1,2		2		
V_{IL}	Low-level input voltage	1, 2		54 Family		0.8
				74 Family		0.8
V_{IK}	Input clamp voltage	3	V_{CC} = MIN, I_I = §			−1.5
V_{OH}	High-level output voltage	1	V_{CC} = MIN, V_{IL} = V_{IL} max, I_{OH} = MAX	54 Family 2.4	3.4	
				74 Family 2.4	3.4	
V_{OL}	Low-level output voltage	2	V_{CC} = MIN, I_{OL} = MAX V_{IH} = 2 V	54 Family	0.2	0.4
				74 Family	0.2	0.4
			I_{OL} = 4 mA	Series 74LS		
I_I	Input current at maximum input voltage	4	V_{CC} = MAX	V_I = 5.5 V		1
				V_I = 7 V		
I_{IH}	High-level input current	4	V_{CC} = MAX	V_{IH} = 2.4 V		40
				V_{IH} = 2.7 V		
I_{IL}	Low-level input current	5	V_{CC} = MAX	V_{IL} = 0.3 V		
				V_{IL} = 0.4 V		−1.6
				V_{IL} = 0.5 V		
O_{OS}	Short-circuit output current*	6	V_{CC} = MAX	54 Family −20		−55
				74 Family −18		−55
I_{CC}	Supply current	7	V_{CC} = MAX			

† For conditions shown as MIN or MAX, use the appropriate value specified under recommended operating conditions.
‡ All typical values are at V_{CC} = 5 V, T_A = 25°C.
§ I_I = −12 mA for SN54'/SN74', −8 mA for SN54H'/SN74H', and −18 mA for SN54LS'/SN74LS' and SN54S'/SN74S'.
* Not more than one output should be shorted at a time, and for SN54H'/SN74H', SN54LS'/SN74LS', and SN54S'/SN74S', duration of short-circuit should not exceed 1 second.

Table 6.2 *continued*

Recommended operating conditions

SERIES 54H SERIES 74H			SERIES 54LS SERIES 74LS			SERIES 54S SERIES 74S			
'H00, 'H04, 'H10, 'H20, 'H30			'LS00, 'LS04, 'LS10, 'LS20, 'LS30			'S00, 'S04, 'S10, 'S20, 'S30, 'S133			
MIN	NOM	MAX	MIN	NOM	MAX	MIN	NOM	MAX	UNIT
4.5	5	5.5	4.5	5	5.5	4.5	5	5.5	V
4.75	5	5.25	4.75	5	5.25	4.75	5	5.25	
		-500			-400			-1000	μA
		-500			-400			-1000	
		20			4			20	mA
		20			8			20	
-65		125	-55		125	-55		125	°C
0		70	0		70	0		70	

SERIES 54H SERIES 74H			SERIES 54LS SERIES 74LS			SERIES 54S SERIES 74S			
'H00, 'H04, 'H10, 'H20, 'H30			'LS00, 'LS04, 'LS10, 'LS20, 'LS30			'S00, 'S04, 'S10, 'S20, 'S30, 'S133			
MIN	TYP‡	MAX	MIN	TYPE‡	MAX	MIN	TYP‡	MAX	UNIT
2			2			2			V
		0.8			0.7			0.8	V
		0.8			0.8			0.8	
		-1.5			-1.5			-1.2	V
2.4	3.5		2.5	3.4		2.5	3.4		V
2.4	3.5		2.7	3.4		2.7	3.4		
	0.2	0.4		0.25	0.4			0.5	
	0.2	0.4		0.25	0.5			0.5	V
					0.4				
		1						1	mA
					0.1				
		50							μA
					20			50	
	-2			-0.4					mA
								-2	
-40		-100	-20		-100	-40		-100	mA
-40		-100	-20		-100	-40		-100	
See table on next page									mA

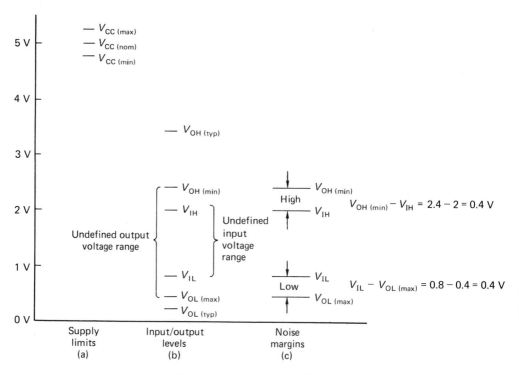

Supply
limits
(a)

Input/output
levels
(b)

Noise
margins
(c)

Fig 6.46 Voltage levels in the 7400 NAND gate

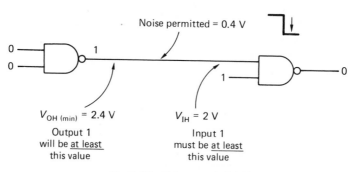

Fig 6.47 Noise margin (high)

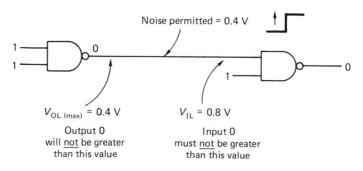

Fig 6.48 Noise margin (low)

Table 6.3 Supply currents. (Courtesy of Texas Instruments Inc.)

	I_{CCH} (mA) Total with outputs high		I_{CCL} (mA) Total with outputs low		I_{CC} (mA) Average per gate (50% duty cycle)
	TYP	MAX	TYP	MAX	TYP
'00	4	8	12	22	2
'04	6	12	18	33	2
'10	3	6	9	16.5	2
'20	2	4	6	11	2
'30	1	2	3	6	2
'H00	10	16.8	26	40	4.5
'H04	16	26	40	58	4.5
'H10	7.5	12.6	19.5	30	4.5
'H20	5	8.4	13	20	4.5
'H30	2.5	4.2	6.5	10	4.5
'LS00	0.8	1.6	2.4	4.4	0.4
'LS04	1.2	2.4	3.6	6.6	0.4
'LS10	0.6	1.2	1.8	3.3	0.4
'LS20	0.4	0.8	1.2	2.2	0.4
'LS30	0.35	0.5	0.6	1.1	0.48
'S00	10	16	20	36	3.75
'S04	15	24	30	54	3.75
'S10	7.5	12	15	27	3.75
'S20	5	8	10	18	3.75
'S30	3	5	5.5	10	4.25
'S133	3	5	5.5	10	4.25

Supply currents (I_{CC}) are given in Table 6.3.
The current supplied depends upon the output condition of the gate. If all outputs are high on the 7400, the typical supply current will be 4 mA (for all four gates), and at maximum 8 mA. If all outputs are low, the typical supply current is 12 mA and the maximum 22 mA. The average typical supply current is $(4 + 12)/2 = 8$ mA for the entire IC, or 2 mA per gate. With this value of I_{CC}, the power consumption will be $5 V \times 2 mA = 10 mW$ for each gate.

6.15 Problems

6.1 Describe how a transistor can operate as a switch. Illustrate your answer with suitable diagrams.

6.2 State typical output voltages for logic 0 and logic 1 levels for a bipolar transistor NOT gate.

6.3 Why is it necessary to use transistors in logic gate circuits?

6.4 Explain the difference between saturating and non-saturating logic.

6.5 Explain the meaning of 'noise imunity', and state why it is necessary and how it is achieved in diode transistor logic gates.

6.6 Sketch the circuit of a DTL NAND gate and explain its operation.

6.7 Sketch the circuit of TTL NAND gate and explain its operation.

6.8 Why do TTL gates have higher maximum operating frequencies than DTL?

6.9 Explain the function of a totem-pole output state in TTL circuits and describe the role of the active load.

6.10 What is meant by fan-in and fan-out in logic gates? Explain how fan-out can be calculated for a TTL gate.

6.11 How do (a) low power and (b) high speed TTL gates differ from standard TTL?

6.12 Explain how you would use an open-collector NAND gate.

6.13 What is the purpose of a tristate device?

6.14 How can a TTL NAND gate be converted to operate with a third high impedance state?

6.15 How can a bipolar transistor be modified to produce a Schottky transistor?

6.16 What feature of transistor operation is improved by the modification of Problem 6.15? Explain precisely why.

6.17 Sketch the circuit of (a) a CMOS NAND gate and (b) a CMOS NOR gate.

6.18 Compare the relative advantages and disadvantages of CMOS logic and TTL logic.

6.19 Explain the operation of the emitter-coupled logic OR/NOR gate and illustrate this with a sketch.

6.20 Produce a voltage level diagram similar to the one in Fig. 6.46 for the Schottky TTL hex inverter 74S04, extracting information from Table 6.2.

6.21 Determine the noise immunity (high and low) for the 74LS20 from the data in Table 6.2.

6.22 What is the fan-out of the 74H20? (See Table 6.2.)

6.23 Calculate the typical power consumption of the 74S133 by referring to Table 6.3.

7

BISTABLE DEVICES

7.1 Introduction

It was stressed in the previous chapter that an important aim in circuit design is the minimisation of delays, as this will allow higher frequency of operation. In combinational logic circuits where gates are used to produce an output as soon as possible after the application of inputs, this is a sensible approach. In sequential logic circuits, however, there is a need to produce delays of varying length from the very short to the very long, not in a haphazard way as with propagation delays, but in an exact way so that accurate timing is possible. Looked at in another way, combinational logic does not involve any storage, whereas sequential logic does. The need for storage of information is vital in digital computers and microprocessors and in digital equipment generally. For example, in 8-bit microprocessors, numbers are stored in bytes of 8 bits (see Chapter 2). Each byte is stored in an 8-bit register, which is an array of eight single-bit storage devices or bistables. Bistables or flip-flops are extremely useful, and find application in one form or another in counters, registers and many other devices. The word 'bistable' implies that the device has two states, i.e. logic 1 and logic 0, and is stable in either of these states. This means that if a bistable is instructed to hold a logic 1 it will do so until told otherwise, and similarly for a logic 0. A bistable can be used therefore as a 1-bit memory element. These elements can be linked together to form groups of 4, 8, 16, or any desired wordlength. Although the concept can be extended to include all memories or semiconductor stores, this chapter is really concerned with the bistable as implemented in a great variety of circuits involved in temporary data storage and control.

7.2 NAND gate bistable (NAND latch)

A bistable can be formed by interconnecting two NAND gates as shown in Fig. 7.1. To examine the operation of the circuit it is necessary to assume an output condition and then apply all possible inputs to check the effect on the outputs.

Condition 1
If, initially,

$$A = 0, B = 0, C = 0$$

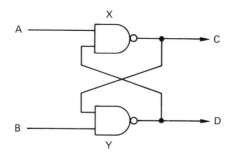

Fig 7.1   NAND latch

then C will become 1 and D will become 1. The assumption is made here that the output C equals 0 *before* the A = B = 0 inputs are applied. Both C and D outputs must go to 1, because 0 is the forcing function for a NAND gate. The inputs to the gates fed back from C and D are irrelevant in this case because of this.

If C_n is the output on C before A and B are applied, and C_{n+1} is the output on C after they are applied then the process can be represented by Table 7.1.

Table 7.1

A	B	C_n	C_{n+1}
0	0	0	→1

The effect on the D output is shown in Table 7.2.

Table 7.2

A	B	D_n	D_{n+1}
0	0	1	→1

Note that D_n must have been 1 if C_n was 0 because of the 0 forcing function being fed back to gate Y.

Bistables usually are designed with complementary outputs, so that one output is the inverse of the other at all times. In the case just explained, $C_{n+1} = 1 = D_{n+1}$ which is not acceptable. For this reason, with the bistable shown the input condition A = B = 0 must be avoided.

Condition 2
If C_n = 0 and the input conditions A = 0, B = 1 are applied then:

(1) As A = 0 (forcing function), C must go to 1 (C→1)
(2) This 1 is fed back to the input of gate Y to join the other input B = 1.
(3) A 1,1 input on a NAND gate gives a 0 output, so D does to 0 (D→0).
(4) This 0 is fed back to gate X input to reinforce the A=0 input and maintain C at 1.

To summarise:

$$A=0 \text{ so } C\to 1$$
$$C\to 1 \text{ so } D\to 0$$
$$D\to 0 \text{ maintaining } C=1$$

and the cycle of events stops. This is illustrated in Table 7.3.

Table 7.3

A	B	C_n	C_{n+1}
0	1	0	$\to 1$

Note that in this case

$$D_n \to 0$$

so that $D_{n+1} = \overline{C}_{n+1}$ as required.

Condition 3

If $C_n = 0$ and the input conditions A = 1, B = 0 are applied then:

(1) As B = 0, D stays at 1.
(2) This 1 is fed back to the input of gate X to join the other input A = 1.
(3) This maintains 0 output on C which is fed back to gate Y.
(4) D stays at 1.

To summarise, C stays at 0, D stays at 1. This is shown in Table 7.4.

Table 7.4

A	B	C_n	D_n	C_{n+1}	D_{n+1}
0	0	0	1	0	1

Condition 4

If $C_n = 0$ and the input conditions A = 1, B = 1 are applied then:

(1) C (= 0) is fed back to the gate Y to join the input B = 1.
(2) D stays at 1.
(3) A = 1 and D = 1 so C stays at 0.

This is shown in Table 7.5.

Table 7.5

A	B	C_n	D_n	C_{n+1}	D_{n+1}
0	0	0	1	0	1

Table 7.6

	A	B	C_n	C_{n+1}
(1)	0	0	0	1
(2)	0	1	0	1
(3)	1	0	0	0
(4)	1	1	0	0

Four conditions have been considered so far, all assuming initially that C = 0, D = 1. These are summarised in Table 7.6.

So far it is difficult to make sense out of what has happened. If the process is repeated, however, with C = 1 initially instead of 0, the results can be predicted from the symmetry of the circuit. This is illustrated in Table 7.7.

Table 7.7

	A	B	C_n	C_{n+1}
(5)	0	0	1	1
(6)	0	1	1	1
(7)	1	0	1	0
(8)	1	1	1	1

A complete truth table is given in Table 7.8.

Table 7.8

A	B	C_n	C_{n+1}	
0	0	0	1	
0	0	1	1	Not used
0	1	0	1	
0	1	1	1	SET
1	0	0	0	
1	0	1	0	RESET
1	1	0	0	
1	1	1	1	No change

The truth table can be divided into four parts:

(1) With inputs A and B both zero, the outputs C and D go to 1. This condition is illegal and not used. (Note that if A = B = 0 and both A and B try to change to 1 at the same time, the output condition is indeterminite as a race will occur. This must not be allowed to happen).

(2) With A = 0, B = 1 the output C goes to 1. This is the SET condition as C is set to 1. Because of this the input B is called the SET (S) input as it equals 1 when C is set to 1.

(3) With A = 1, B = 0 the output C goes to 0. Input A is called the RESET (R) input.
(4) With A = 1, B = 1, if C is 0 (and D = 1), initially C will stay at 0. If C is 1 (and D = 0) C will stay at 1. This is called the 'no change' condition.

The output C is usually represented by the symbol Q and the output D by $\overline{Q}$. The diagram is redrawn in Fig. 7.2.

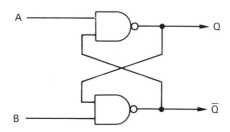

Fig 7.2 NAND latch with Q and $\overline{Q}$ outputs

The truth table for the NAND latch is given in Table 7.9.

Table 7.9

S	R	Q_{n+1}	
0	0	X	Not used
0	1	0	RESET
1	0	1	SET
1	1	Qn	No change

The use of Q and $\overline{Q}$ to represent outputs is universal and applied to all bistables.

Although the mechanics of the operation of the latch using two NAND gates has been explained in some detail, the significance of all this has not. To appreciate this it is necessary to refer back to the original explanation of what a bistable is. A bistable was defined as having two stable states—so how does this apply in the case of the NAND latch? One way of approaching the problem is to consider the practical implementation of the latch using a 7400 TTL NAND integrated circuit as illustrated in Fig. 7.3.

Note that if A and B are not connected they will float 'high' to a logic 1 which is equivalent to the no change condition. If a 1 is applied to B and a 0 is applied to A (SET condition) Q will go to 1 and $\overline{Q}$ to 0.

If A is now changed from 0 to 1, or if both inputs A and B are disconnected, the effective A = B = 1 inputs (no change) will maintain the bistable at logic 1. Thus a 1 has been 'latched' into, or 'stored' in, the bistable.

If a 0 is applied to B and a 1 to A (RESET condition), Q will go to 0 and $\overline{Q}$ to 1. If B is now changed from 0 to 1, or A and B are disconnected, the logic 0 is maintained on Q. Thus a 0 has been stored in the bistable. The criterion that a bistable should have two stable states has been satisfied.

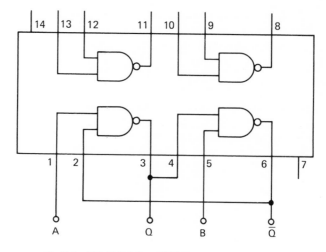

Fig 7.3 TTL 7400 Quad NAND connected as a latch

7.3 NAND latch applications

A common application of a NAND latch is in debouncing switches.

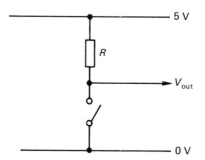

Fig 7.4 Switch without debouncing circuit

Consider the simple switch shown in Fig. 7.4. When the switch is open the output voltage V_{out} is given by $V_{out} = 5$ V. With the switch closed, $V_{out} = 0$ V. The action of closing the switch however may involve a sequence of closures caused by contact bounce. This is illustrated in Fig. 7.5 which shows three bounces.

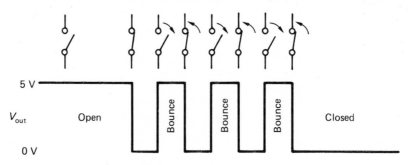

Fig 7.5 Contact bounce

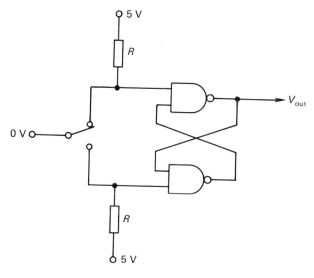

Fig 7.6   Debouncing circuit

For applications involving the use of a switch as a pulse generator this is fatal, as several extra pulses will be generated. A switch can be debounced by using a NAND latch as shown in Fig. 7.6.

7.4 Unclocked and clocked SR bistables

Allowing inputs to float high as described in the example in section 7.2 is not a satisfactory method of latching information into a bistable. A design of bistable that performs the latching by a more positive method is introduced in Fig. 7.7.

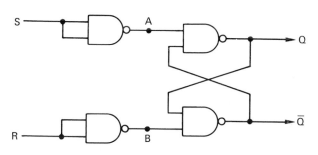

Fig 7.7   Unclocked SR bistable

Two additional NAND gates, connected as inverters, have been added. Input S is called the SET input and R is called the RESET input. The truth table for this arrangement is given in Table 7.10. Note that if S = R = 0, then the no change condition exists; if S = R = 1, the no use condition exists.

Alternatively the truth table can be abbreviated, as shown in Table 7.11.

If the circuit of Fig. 7.7 is modified to include a control input, the result is as shown in Fig. 7.8.

Table 7.10

S	R	Q_n	Q_{n+1}	
0	0	0	0	
0	0	1	1	No change
0	1	0	0	
0	1	1	0	RESET
1	0	0	1	
1	0	1	1	SET
1	1	0	X	
1	1	1	X	No use

Table 7.11

S	R	Q_{n+1}	
0	0	Q_n	No change
0	1	0	RESET
1	0	1	SET
1	1	X	No use

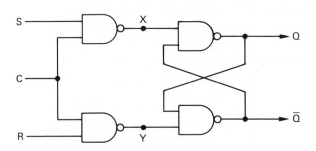

Fig 7.8   Clocked SR bistable

If the control input, C, is set to logic 1 then the input gates will operate as inverters. The circuit then becomes identical to that of Fig. 7.7.

If the control input is set to logic 0 then, again, because of the forcing function concept, points X and Y in Fig. 7.8 will become logic 1. This is equivalent to the no change condition specified for the NAND latch, so that the Q and $\overline{Q}$ outputs will be held at whatever value they had before the transition of the control inputs from 1 to 0. With C at 0 the inputs S and R are completely ignored by the bistable, so that it stores the output Q until told to do otherwise. Changing C from 0 to 1 at any instant will immediately (ignoring propagation delays) produce the output appropriate to the SR input values. The bistable in Figure 7.8 is represented in block diagram form as in Figure 7.9.

The input C is more commonly referred to as a **clock input**.

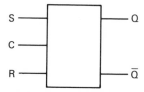

Fig 7.9   SR bistable symbol

7.5 Master–slave SR bistable

As a single bit storage device, the simple SR bistable described is adequate. It will store a 0 or a 1 if appropriate input conditions are applied. This bistable, however, has two major faults that limit its applications. These are:

(1) The bistable cannot be used as part of a serial register to store word.
(2) The S = R = 1 condition does not perform a useful function.

A serial register requires a common clock input to a number of bistables, as shown in Fig. 7.10

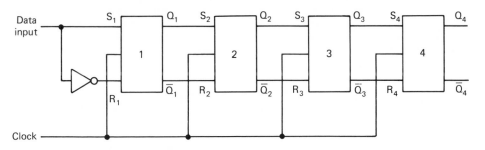

Fig 7.10   Serial register

In order to ensure that only SET and RESET conditions are encountered, the Q outputs of the bistables are connected to the S inputs of the next bistable in sequence and the $\overline{Q}$ outputs are connected to R inputs of the next bistable. Clearly this is not possible on the first stage, so to force S_1 and R_1 to be complementary, an inverter is added as shown. This leaves a single data line to act as an input to the register. The idea is that by setting the data line (for example) to a 0, an initial clock change from 0 to 1 will allow the data to be stored in bistable 1 and wait until the next clock pulse before going through to the second stage. Unfortunately, as soon as the logic 0 appears at the output of bistable 1, it instantly becomes the input to bistable 2 which *also* has a high clock input. The 0 is therefore stored in bistable 2 and similarly bistables 3 and 4. Putting a 1 on the data line does not help; logic 1's will appear on Q_1, Q_2, Q_3 and Q_4 as soon as the clock goes high.

What is needed is some method of preventing the output of Q_1 from reaching S_2 (and of course $\overline{Q}_1$ from reaching R_2) until the clock input to bistable 2 has been reduced to zero, which would inhibit any further transmission of data. If this is applied at each stage, then a bit pattern can be stored by feeding in one bit at a time. If the bistable is modified by the addition of an identical second stage, this can be achieved as shown in Fig. 7.11.

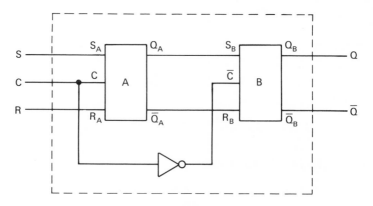

Fig 7.11   SR master–slave bistable

There are two bistables, A and B, which combine to form a single bistable. A is the **master** and B is the **slave**. An inverter is added so that when the clock is high on A it is low on B and vice versa. This master – slave bistable will allow data to be transferred through to Q_A as long as the clock input C is high (*not* merely during the positive edge of the clock). The data cannot get through to Q_B, the composite bistable output, until $\bar{C} = 1$, i.e. when the input clock C goes low. If this master–slave bistable is followed by another master–slave bistable as part of a register system, this means that the data cannot go through the first bistable until the clock is low again, i.e. a complete clock pulse is required to transfer the data from the SR inputs to the Q output. When the clock has gone low the next bistable is inhibited, solving the problem of data rippling through each stage of the register. This is illustrated in Fig. 7.12.

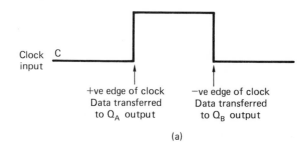

(a)

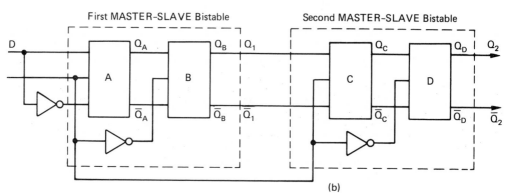

(b)

Fig 7.12   (a) Timing on master–slave bistable; (b) Master–slave bistable operation

Figure 7.13 illustrates the process of transferring a logic 1 from the SET input to the Q output of a master–slave bistable.

In NAND gates the master–slave (M–S) bistable can be implemented as shown in Fig. 7.14

Although the SR M–S flip-flop solves the problem of ripple through in shift registers, it does not solve the problem of the S = R = 1 condition. The addition of a clock to the

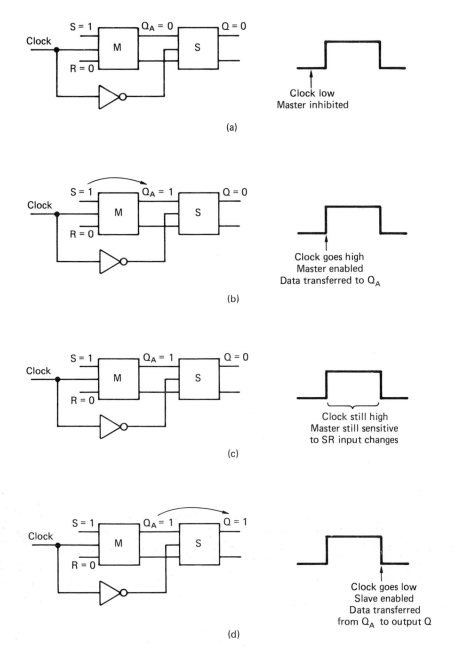

Fig 7.13 Data transfer in a SR master–slave bistable

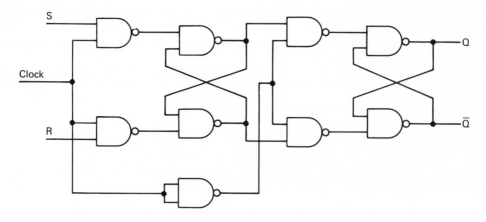

Fig 7.14   NAND gate implementation of SR master–slave bistable

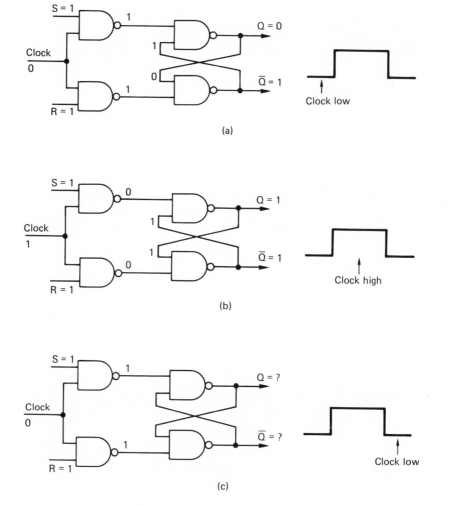

Fig 7.15   Race condition in clocked SR bistable

simple SR flip-flop as in Fig. 7.8 modifies the 'no use' condition. The sequence of events shown in Fig. 7.15 illustrates this.

In Fig. 7.15 (c), the final levels of Q and $\overline{Q}$ cannot be determined although one output must go to 0 and the other to 1.

As soon as the clock goes low, the inputs to the NAND latch go high. Both latch gates have 1, 1 inputs so both outputs aim for logic 0. One gate will be fractionally quicker than the other, and so will reach the 0 level first. This 0 will be fed back to the other gate, forcing its output to 1 and latching the flip-flop. In other words unlike the previous no use condition of the unclocked RS flip-flop, the clocked RS flip-flop has an **indeterminate** state for S = R = 1 inputs.

7.6 Master–slave JK bistable

In order to eliminate the indeterminate state, the SR master–slave flip-flop can be modified quite simply to provide an extremely useful condition. The circuit of Fig. 7.14 is altered as shown in Fig. 7.16 to produce a **master–slave JK** flip-flop.

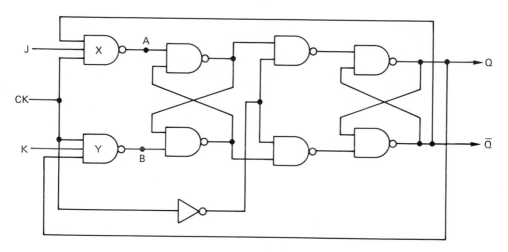

Fig 7.16   JK master–slave bistable

If it is assumed that J (S) = K (R) = 1 and that Q = 0, $\overline{Q}$ = 1 initially, then Q is fed back to the input of gate Y and $\overline{Q}$ to the input of gate X. The 0 input from Q to gate Y, however, forces the output of Y to logic 1. The flip-flop 'thinks' of this as being the same as a 0 input on K (R) and so follows the SET condition, putting Q = 1 and $\overline{Q}$ = 0. If the process is repeated, the 0 from $\overline{Q}$ is fed back to gate X which produces the equivalent to the J (S) = 0, K (R) = 1 or RESET condition, so that Q goes to 0 and $\overline{Q}$ goes to 1. The overall effect is therefore that the flip-flop **toggles**, changing state on the negative edges of the clock. The truth table of the JK flip-flop is given in Table 7.12.

Table 7.12

J	K	Q_{n+1}
0	0	Q_n
0	1	0
1	0	1
1	1	$\overline{Q}_n$

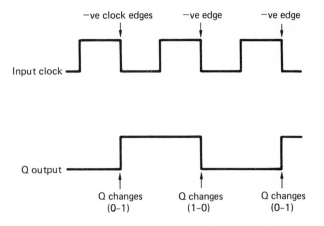

Fig 7.17   Frequency division using JK-MS bistable

If a continuous clock (square wave) is fed into the bistable with $J = K = 1$, the resulting waveform on Q is as shown in Fig. 7.17.

The resulting waveform is half the frequency of the incoming waveform so that if the clock frequency was 100 kHz the output frequency on Q would be 50 kHz. This is an extremely useful feature which is widely exploited.

An example of the use of a single bistable is in the conversion of a waveform with an unequal mark-to-space ratio into one (at half frequency) with an equal mark-to-space

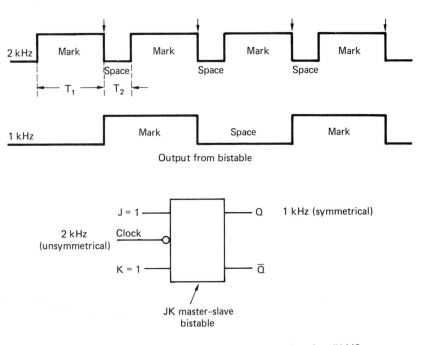

Fig 7.18   Production of equal mark-to-space ratio using JK-MS

ratio. The mark-to-space ratio or pulse duty factor is the ratio of the time for which a repetitive waveform is in a high (mark) state (T_1) to the time in the low (space) state (T_2), i.e.

$$\text{Pulse Duty Factor} = \frac{T_1}{T_2}$$

An equal mark-to-space ratio has $T_1 = T_2$. A number of commercially available integrated circuit timer/oscillators produce waveforms which need this conversion process. As an example, assume that a 555 timer chip produces an unsymmetrical waveform as shown in Fig. 7.18.

Note that the circle (inversion) on the clock input denotes operation on the 1 to 0 transition of the clock. If a 1 kHz symmetrical waveform is required, then suitable component values can be chosen in the frequency setting of the timer circuit to produce twice the required frequency at the output of the timer chip. Dividing the unsymmetrical waveform by 2 with a JKMS flip-flop will provide the correct frequency with an equal mark-to-space ratio.

7.7 Preset and clear inputs

The NAND latch can be modified as shown in Fig. 7.19.

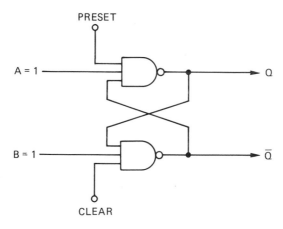

Fig 7.19   Preset and clear inputs on NAND latch

A 0 applied to the PRESET input will force Q to 1 and $\overline{Q}$ to 0. A 0 applied to the CLEAR input will force $\overline{Q}$ to 1 and Q to 0. It is essential to avoid setting both PRESET and CLEAR to 0, otherwise Q will equal 1 and $\overline{Q}$ will equal 1. The Q and $\overline{Q}$ states introduced by the PRESET or CLEAR inputs will be latched in and held. The same idea can be applied in the JK master-slave flip-flop, as shown in Fig. 7.20.

The JK inputs only become effective upon receipt of a clock pulse and are therefore referred to as **synchronous** inputs.

The PRESET and CLEAR do not depend upon the clock and are therefore said to be **asynchronous**.

7.8 A TTL JK master-slave bistable

A typical JK master-slave bistable is the 7476, shown in Fig. 7.21.

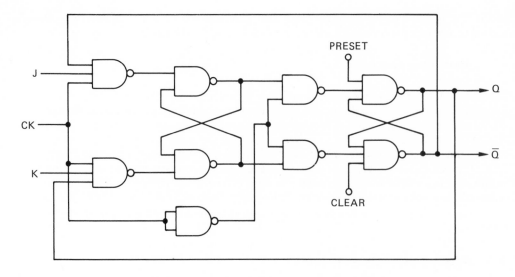

Fig 7.20   Preset and clear inputs on JK-MS bistable

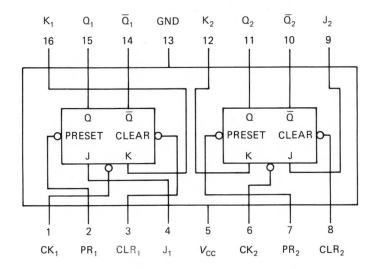

Fig 7.21   7476 bistable

It is in fact a dual JK and has **PRESET** and **CLEAR** inputs which act as described in Fig. 7.20 in that they require a logic 0 to operate, i.e. that they are 'active low'. This is indicated by the small circle on the **PRESET** and **CLEAR** inputs.

7.9 D-type bistable

Data or **Delay** (D-type) flip-flops have a single input line (D), instead of two input lines as in the case of the SR and JK bistables. D-types exist in a variety of forms. One method of producing a D-type is shown in Fig. 7.22.

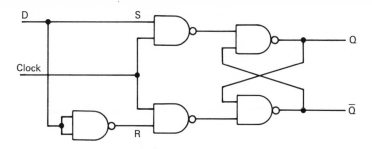

Fig 7.22 D-type bistable from clocked SR

The addition of an inverter in the RESET line, as shown, eliminates the possibility of the S = R = 0 or S = R = 1 inputs occurring, as indicated in the modified truth table given in Table 7.13.

Table 7.13

S	R	Q_{n+1}
0	1	0
1	0	1

In other words, the bistable will SET or RESET only, setting Q to 1 if D = 1 and resetting Q to 0 if D = 0. The D-type shown is identical to the first stage of the (unsuccessful) register shown in Fig. 7.10. Whatever input is applied to D will be transferred to the Q output as soon as the clock goes high. When the clock goes low again the data will be latched. The truth table for D-type is given in Table 7.14.

Table 7.14

D	Q_{n+1}
0	0
1	1

The symbol for D-type is given in Fig. 7.23.

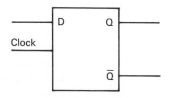

Fig 7.23 D-type symbol

The D-type bistable can be used to divide a clock input frequency by two, by connecting the $\overline{Q}$ output to the D input as shown in Fig. 7.24.

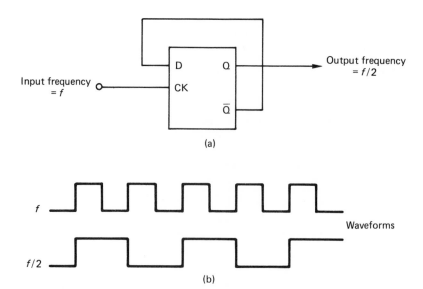

(a)

(b)

Fig 7.24   Dividing frequency by 2-with a D-type

A JK master-slave flip-flop can also be converted to a D-type by adding an inverter. This is illustrated in Fig. 7.25.

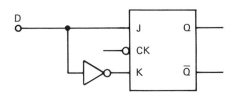

Fig 7.25   D-type from JKMS

The D-type of Fig. 7.25 has the same truth table as that of Figure 7.24, but operates at a different point on the clock waveform. The former operates on the positive edge of the clock and the latter operates on the negative edge. The positive edge is sometimes known as the **leading edge**, and the negative edge as the **trailing edge**. In both cases, data is admitted on the high level of the clock. The D-type in Fig. 7.24 also sets the output of the flip-flop to 0 or 1 as soon as the level on the clock is high. The D-type of Fig. 7.25 only permits it to be sent to the output when the clock level goes low. Both of these are referred to as **level-triggered** flip-flops. Alternatives to these methods exist – not only for D-types but for all types of bistables. These are generally known as **edge-triggered** bistables.

An edge-triggered bistable is only sensitive to input values during the period in which the clock is changing from one level to another. The transition during which the bistable is enabled can be from 0 to 1 (positive-edge) or from 1 to 0 (negative-edge).

An example of the operation of a positive edge triggered D-type bistable is given in Fig. 7.26.

Note that the D input is sampled on the positive edge of the clock only, at time t_1 when D = 1. The change in level of D from 1 to 0 at time t_2 will not be accepted.

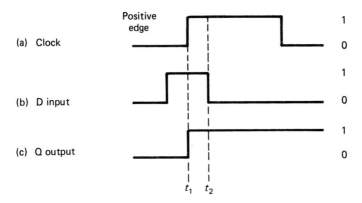

Fig 7.26   Positive edge triggered D-type operation

Figure 7.27 illustrates the operation of a level triggered D-type bistable to enable a comparison to be made.

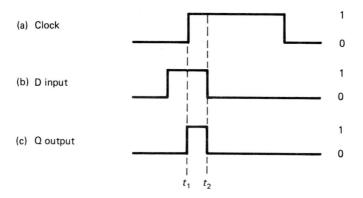

Fig 7.27   Level triggered D-type operation

In this case the Q output will follow the D input as long as the clock level is high, so that in Fig. 7.27 the Q output will go to 0 at time t_2 in response to the 1–0 transition on D.

A positive edge triggered D-type is shown in Fig. 7.28

Assume that initially data = 1, clock = 0, Q = 0.

Step 1: Clock goes from 0 to 1.
Step 2: Output of B goes to 0.
Step 3: Output Q goes to 1, $\bar{Q}$ goes to 0, Q latched.
Step 4: Assume data goes to 0.
Step 5: Output of D goes to 1.

C output is maintained at 1 – no further action.

In other words, the data input is only effective during the positive transition of the clock. At all other times the data is 'locked out'.

The positive edge triggered bistable is given the symbol shown in Fig. 7.29.

A negative edge trigged bistable has the symbol shown in Fig. 7.30

When designing a system using a variety of integrated circuit bistables, registers and

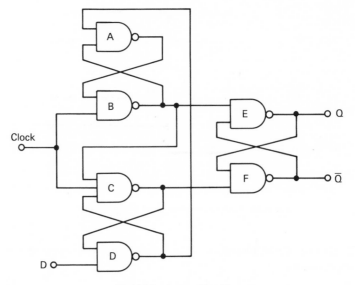

Fig 7.28 D-type bistable

Fig 7.29 Positive edge-triggered D-type symbol   **Fig 7.30** Negative edge-triggered D-type symbol

counters, it is essential to verify the type of triggering of each one. If one part of the system uses level triggered devices and another part of the system uses negative edge triggered circuits, ensure that this is part of the design and not just a careless selection of device types. A typical D-types flip-flop is the 7474 TTL D-type flip-flop illustrated in Fig. 7.31. This is a positive edge triggered device.

7.10 T-type bistable

It is shown earlier in this chapter that it is possible to have the condition in which the output from a flip-flop alternates between 0 and 1 as successive clock pulses are fed in. This was referred to as 'toggling'. A bistable which operates in this way is referred to as a T-type bistable, and is given the symbol shown in Fig. 7.32.

It is possible to have either positive or negative edge triggered, or level triggered, T-type flip-flops.

7.11 CMOS bistables

Many CMOS bistables incorporate a device known as a **transmission gate**. The symbol for a transmission gate is given in Fig. 7.33(a).

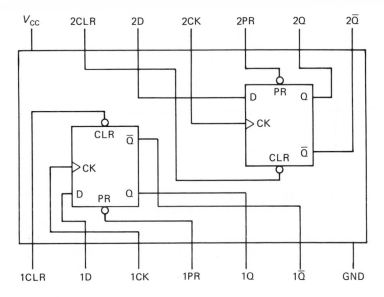

Fig 7.31 7474 D-type

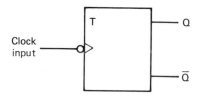

Fig 7.32 Symbol for T-type bistable

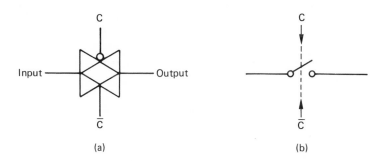

Fig 7.33 Transmission gate

The device is equivalent to the switch shown in Fig. 7.33(b). When the control input C is at logic 0, the switch is closed and the output follows the input. Application of a logic 1 to control input C will open the switch. If the symbol is shown with C and $\bar{C}$ reversed as in Fig. 7.34, then application of a logic 0 to the control input will open the switch. Application of a logic 1 to C will close the switch.

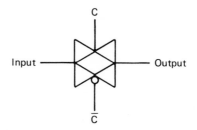

Fig 7.34 Inverted controls

A D-type bistable, the 4013, utilises four of these gates, two of which are the inverted control type. The circuit for the 4013 is given in Fig. 7.35.

Inputs to the device are D (data), S_D (SET direct), C_D (CLEAR direct) and C, the clock input. $\overline{C}$ is generated internally. Outputs are Q and $\overline{Q}$. S_D and C_D are asynchronous inputs and are active high, carrying out the same functions as the PRESET and CLEAR inputs previously described for TTL flip-flops. The clock is fed to the transmission gate inputs via a Schmitt trigger which allows for slow rise and fall times on the clock.

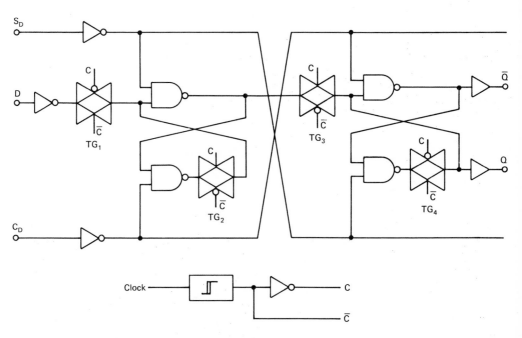

Fig 7.35 D-type flip-flop (4013)

The flip-flop consists of two NAND latches acting as master and slave. In each latch, one of the feedback paths is via a transmission gate, one directly clocked and the other with an inverted clock. This means that at any instant, two transmission gates are active (i.e. short circuits) and two are high impedance (i.e. open circuits). The action of the bistable will be illustrated using switches instead of the transmission gates. Figures 7.36, 7.37 and 7.38 give an example.

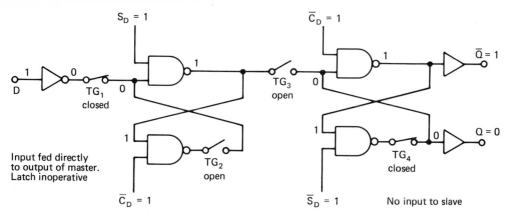

Fig 7.36   D Master–slave before positive edge

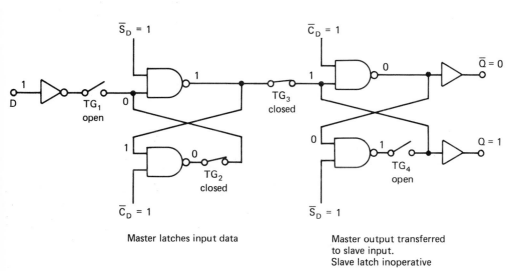

Fig 7.37   D Master–slave after positive edge

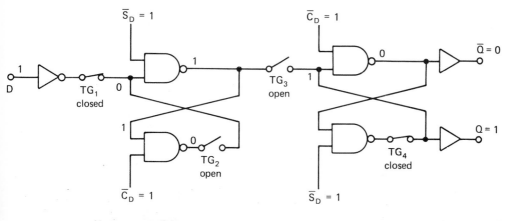

Fig 7.38   D Master–slave after negative edge

In this example, the asynchronous inputs are inoperative so that $S_D = C_D = 0$. The data input is accepted while the clock is low, as can be seen in Fig. 7.36, but will not be latched into the master until the arrival of a positive clock edge. This is because TG_1 is closed, enabling data entry, and TG_2 is open, preventing NAND latch action. While the clock is low, data cannot reach the slave because TG_3 is open. When the clock does go high as shown in Fig. 7.37, the master latches the data by closing TG_2 and isolates itself from the D input by opening TG_1. At the same time, the data is fed to the slave input and arrives at the Q output but is not latched. When the negative clock edge arrives, the slave is isolated from the master and is latched by TG_4 closing; this is illustrated in Fig. 7.38. The timing sequence is shown in Fig. 7.39.

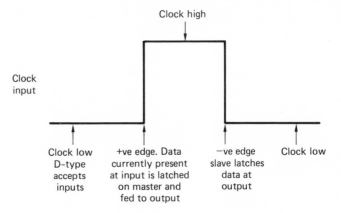

Clock low	+ve edge. Data	−ve edge	Clock low
D–type	currently present	slave latches	
accepts	at input is latched	data at	
inputs	on master and	output	
	fed to output		

Fig 7.39   Timing sequence of D-type flip-flop

Thus data is transferred from D to Q on the positive edge, making this a positive edge triggered master–slave bistable, but is only latched at the negative edge. Data is locked out while the clock is high.

A CMOS JK flip-flop is illustrated in Fig. 7.40.

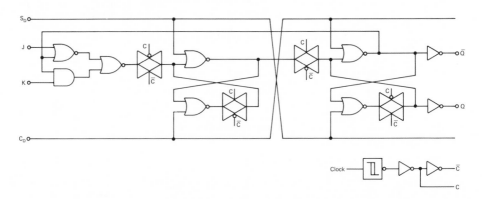

Fig 7.40   CMOS JK flip-flop

The operation of the JK illustrated is very similar to that of the D-type, although of course it follows the conventional JK truth table. It also is a positive edge triggered master – slave device. The JK bistable illustrated is the CMOS 4027.

7.12 Set-up and hold time

Data sheets on bistables will supply values for **set-up time** (t_{su}) and **hold time** (t_{hold}). The set-up time is the minimum time for which the inputs (D, J, K, etc.) must be stable before the triggering edge of the clock occurs. The hold time is the minimum time that these inputs must remain stable after the triggering edge of the clock has arrived. If these times are not taken into account, correct operation of the bistable can not be guaranteed. The set-up and hold times for a positive edge triggered D-type bistable are given in Fig. 7.41. Typical values of t_{su} are 10–50 ns, and t_{hold} 0–10 ns.

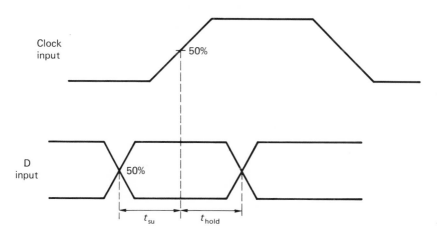

Fig 7.41   Set-up and hold times for a positive edge triggered D-type bistable

7.13 Bistable applications

Bistable devices or flip-flops are widely used in a variety of applications. Typical applications are:

(i)   latches and buffers
(ii)   registers and counters (as described in the following chapters)
(iii) flags (one bit memory elements)
(iv)   frequency dividers

Bistables also play an important role in control and timing circuits. They are available in TTL, CMOS, ECL and other logic families, including HCMOS.

7.14 Problems

7.1   Show that two NOR gates connected in the same way as the NAND gates of Fig. 7.1 will also act as bistable latch.

7.2   Can the NOR latch of problem 7.1 be used to debounce switches? Explain!

7.3   Sketch symbols for the following flip-flops:
(a) JK master–slave.
(b) D-type (level triggered).
(c) SR master–slave.
(d) D-type (edge triggered).

7.4   What would the effect be of setting both PRESET and CLEAR to 1 on a JKMS flip-flop?

7.5   Give four applications of bistables.

7.6   Explain the principle of the master–slave flip-flop, and explain why its need arises.

7.7   Prove the truth table for an unclocked SR bistable using four NAND gates.

7.8   Distinguish between the 'no use' condition and the 'indeterminate' condition as applied to bistables.

7.9   How does a race condition arise in a clocked SR bistable?

7.10   How could JK-MS flip-flops be used to divide a frequency by 4?

7.11   What is the purpose of the PRESET and CLEAR inputs on bistables?

7.12   Explain the operation of the CMOS transmission gate.

7.13   Fig. 7.40 gives the circuit of a CMOS JK flip-flop.
Explain its operation in detail, giving a step by step description of the process.

7.14   Explain the meaning of set-up and hold times with reference to bistable operation.

8

REGISTERS

8.1 Introduction

Chapter 7 deals with the devices that are used to store single bits. This chapter deals with registers – devices that store groups of bits or words. The need for such devices is established in Chapter 2, where arithmetic operations involve the storage and manipulation of binary numbers of various wordlength.

A register is constructed from N flip-flops, where N is the word length to be stored. Provision must be made to store bit patterns in the register and to read these bit patterns when required. Data entry and retrieval can either be serial, with the resulting loss of speed, or parallel, with increased hardware requirement.

Often there is a requirement to shift bit patterns that are stored in a register to the left or to the right, as demonstrated in Chapter 2 when multiplication and division were discussed. In this case a **shift register** is needed.

Registers can be constructed from separate flip-flops, or they can be obtained in various integrated circuit forms. They are component parts of many medium and large scale integrated circuits, and can be implemented using a variety of logic families. This chapter will describe the construction, operation and application of the various forms of register.

8.2 Register input/output techniques

The methods of input and output can be represented by four register types:

(i)   SISO: Serial Input Serial Output.
(ii)   SIPO: Serial Input Parallel Output.
(iii) PIPO: Parallel Input Parallel Output.
(iv)   PISO: Parallel Input Serial Output.

Many registers have shift right facility. Some registers have shift left facility, and a number of registers incorporate bidirectional operation, sometimes referred to as **reversible** shift registers.

In general, registers can be represented as rectangular boxes subdivided into a number of individual cells of binary units, provided that the wordlength is not too great. For a large number of bits, the subdivisions are usually omitted. The least significant and the most significant digits are marked, to indicate the wordlength, as shown in Fig. 8.1.

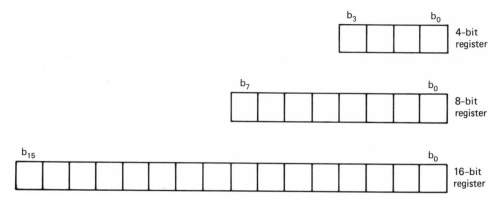

Fig 8.1 Block diagram representation of the registers

Some of the different types of registers are represented diagrammatically in Fig. 8.2. A block diagram of a serial-in serial-out, SISO, register is shown in Fig. 8.2(a). The serial-in serial-out and parallel-out, SISO/PO, arrangement is shown in Fig. 8.2(b), while the parallel-in parallel-out is shown in Fig. 8.2(c) and a bidirectional universal shift register is represented by Fig. 8.2(d).

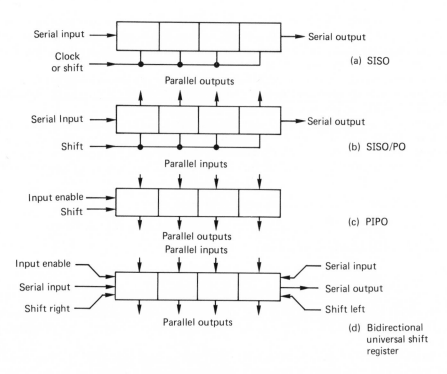

Fig 8.2 Diagrammatic representation of shift registers

8.3 Labelling of individual stages of registers and counters

The convention for labelling the individual stages of flip-flops, registers and counters will be considered first, before dealing with the bulk of the theory in this chapter.

There is no recognised standard for marking the individual stages, and this is reflected in the methods used in most publications, and in the technical data published by manufacturers. When practical applications are considered, however, it is convenient to represent the least significant bit at the extreme right hand side of the diagram, so that it corresponds to conventional written representation of binary numbers. It is for this reason that both registers and counters (described in the next chapter) are labelled, whenever possible, with the least significant digital (LSD) situated on the right hand side, as shown in Fig. 8.3.

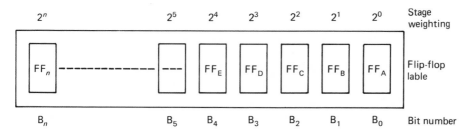

Fig 8.3   Labelling of the individual stages

8.4 Serial transmission

In serial transmission, the specific binary information is represented as a train of pulses, the presence of a pulse represents bit 1 and its absence represents bit 0.

Example 8.1
An example of serial operation is shown in Fig. 8.4.

Fig 8.4   Serial shift registers

Two four-bit registers are illustrated here. One represents register A and the other register B. The register B initially contains 1000, and register A holds 1101, which is to be transferred to register B. To accomplish the transfer, four clock, or shift, pulses are required. As the pulses are applied, the information from register A is shifted to register B one bit at a time. The data previously held in register B is pushed out and lost. The sequential transition of the data is shown in Fig. 8.5.

The content of register A, after each pulse, is progressively replaced with 0's. The speed with which the information is transmitted depends on the number of bits to be transferred. For a large wordlength, therefore, serial data transmission is slow.

8.5 Serial input, serial output, shift register

A basic four-bit shift register can be constructed using four D-type flip-flops. A suitable circuit is shown in Fig. 8.6.

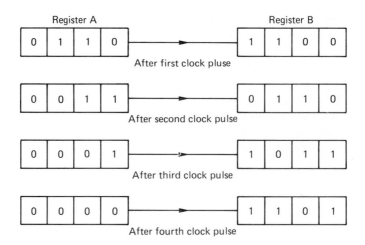

Fig 8.5   Transfer of data sequence

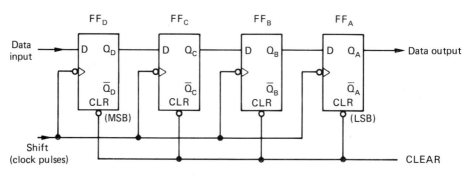

Fig 8.6   SISO shift register

The operation of the circuit is as follows. The register is first cleared by an application of a '0' to the clear line, forcing all four outputs to go to zero. The input data is then applied sequentially to the D input of the first flip-flop on the left, FF_D, and is transmitted one bit at a time, left to right, during each clock pulse. The transition of data through the register is shown in Table 8.1, assuming a data word to be 1101. The least significant bit of the data has to be shifted through the register from FF_D to FF_A.

Table 8.1   Outputs of the shift register

Clock pulse	Data input		Output at Q of each register after a clock pulse			
			Q_D	Q_C	Q_B	Q_A
1	1	→	1	0	0	0
2	0	→	0	1	0	0
3	1	→	1	0	1	0
4	1	→	1	1	0	1

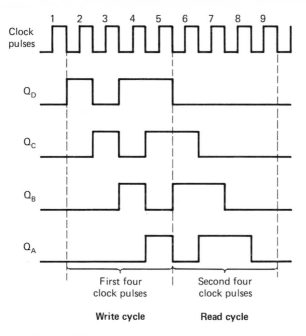

Fig 8.7 Waveform transitions for SISO shift register

The idealised waveforms for the transition are shown in Fig. 8.7 and the operation is as follows.

With the least significant bit of the data applied to the flip-flop D, FF_D, the output of Q_D changes its state from 0 to 1 on the falling edge of the clock pulse. All other outputs remain at 0.

On the second clock pulse, the 1 from Q_D is transferred to Q_C, and Q_D changes to 0, since the second data bit is a 0.

On the third clock pulse, the 1 from Q_C is transferred to Q_B, the 0 from Q_D is transferred to Q_C, and a new bit 1 entered, causing Q_D to change to a 1.

Finally, on the fourth clock pulse, the data is shifted once more and the last, i.e. the most significant, bit of the data is then entered to FF_D, setting Q_D at 1.

This shows that the data is shifted to the right by one flip-flop on receipt of each clock pulse.

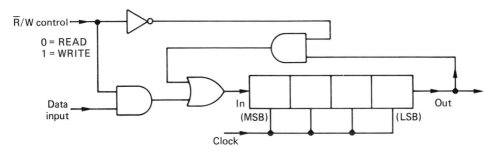

Fig 8.8 Non-destructive read-out arrangement

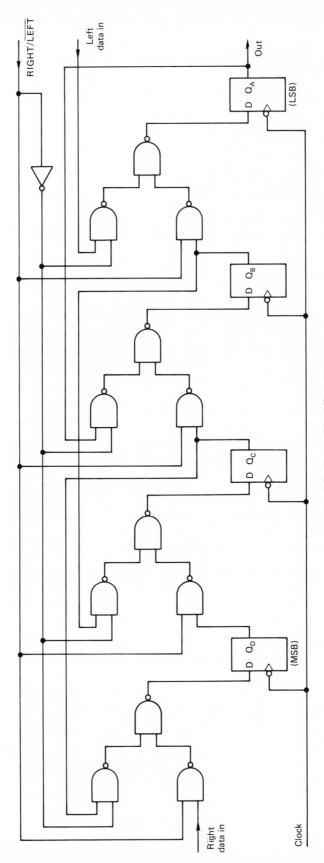

Fig 8.9 Reversible shift register

In order to **read** the data from the series shift register, the contents can be shifted out or transferred to another register with the data input held at 0, i.e. resetting the register at the same time. This results in a **destructive readout**, i.e. the original data is lost, and at the end of the read cycle (Fig. 8.7) all flip-flops are reset to zero. To avoid the loss of data, an arrangement for a non-destructive reading can be incorporated in the design. This is done by adding a suitable gating system, as shown in Fig. 8.8, for recirculating the data.

Two AND gates, an OR gate and an inverter are used to perform the operation. The data is loaded to the register when the control line is at logic 1, i.e. WRITE = 1. The data can be shifted out of the register when the control line is at 0, i.e. READ = 0.

8.6 Bidirectional shift registers

The shift right operation, discussed above has the effect of successively dividing the binary number by two. If the operation is reversed, i.e. the binary number is shifted once to the left, this has the effect of multiplying the number by two. With suitable gating arrangement a serial shift register can perform both operations.

Example 8.2
A bidirectional, or reversible, shift register using D-type flip-flops is shown in Fig. 8.9.

Here a set of NAND gates are configured as OR gates to select data inputs from the right or left adjacent bistables, as selected by a control line. The register can be cleared by holding the data inputs at zero and applying four shift right pulses, or four shift left pulses.

The length of the register can be adjusted to hold any number of binary digits by adding the required number of flip-flops together with the three NAND gates per stage.

Example 8.3
Shift register using SR or JK flip-flops
A basic four-bit forward shift register is shown in Fig. 8.10. Generally, either SR or JK flip-flops can be employed.

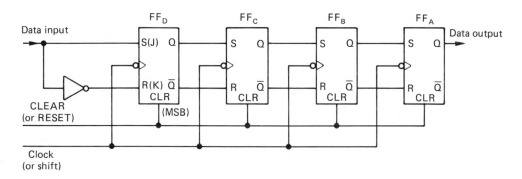

Fig 8.10   Shift register using SR or JK flip-flops

Flip-flop FF_A holds the least significant bit and flip-flop FF_D holds the most significant bit.

The first stage, FF_D, is fed with complementary signals, i.e. the stage operates as a D-type flip-flop. In order to set a 1 in FF_D, S (or J) has to be high and R (or K) has to be low for the duration of the first clock pulse. From there on, direct connections are made from the Q output of one flip-flop to the S input of the next stage. Similarly, $\overline{Q}$ outputs are

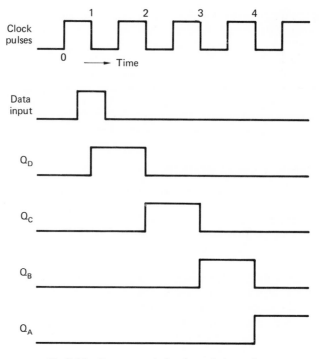

Fig 8.11   Data transmission through the register

connected to R inputs throughout. Now, assume that all Q outputs are reset to zero. With the data input held low, $Q_D = 0$ and $\overline{Q}_D = 1$; also $Q_C = Q_B = Q_A = 0$ and $\overline{Q}_C = \overline{Q}_B = \overline{Q}_A = 1$, irrespective of the clock. If the data input is taken to a logic 1, Q_D will go high at the end of the clock pulse. If the data input is then held low for the duration of the next three clock pulses, the 1 entered during the first clock pulse will be shifted through the register to Q_A at the end of the fourth clock pulse, as illustrated in Fig. 8.11.

Note that in Fig. 8.10 the flip-flops are of the master–slave type, and they become set after the completion of the clock pulse. In fact, all four stages of the shift register could be connected as D-type flip-flops without affecting the order of data transmission. Shift registers, as seen previously, can be designed using flip-flops made up of cross coupled gates in the master–slave configuration, or using JK flip-flops in integrated circuit form. However, the most common practice is to employ a readily available register chip. In general, the shift registers available in the TTL 7400 series use SR flip-flops, and these available in the CMOS 4000 range use D-type circuits.

Example 8.4
Serial-input serial-output shift register
An example of an eight-bit serial-input serial-output shift register is the 7491 MSI circuit composed of eight SR master–slave flip-flops.

Single line data and input controls are gated via a NAND gate. An inverter is used to form the complementary inputs to the first stage of the register. The SR flip-flops are of the negative edge triggered variety. The clock input is inverted to allow a positive edge triggering facility for data shifting.

A functional block diagram of the shift register is shown in Fig. 8.12.

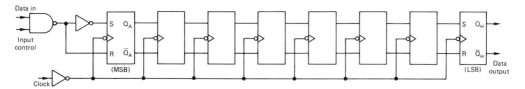

Data in

Input control

Clock

(MSB)

(LSB)

Data output

Fig 8.12 Eight-bit serial input register

The circuit function as a shift right register only, but provides two outputs: the data output Q_H and an inverted version of the data at $\overline{Q_H}$.

8.7 Parallel transmission

In parallel transmission the binary information is transferred simultaneously on receipt of an enabling pulse. The system is illustrated in Fig. 8.13. The rectangular boxes represent individual flip-flops making up a register.

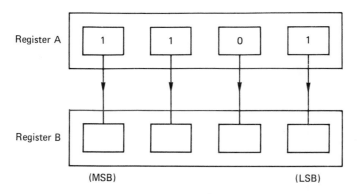

Register A

1   1   0   1

Register B

(MSB)   (LSB)

Fig 8.13 Block schematic of parallel transmission

The speed of operation in parallel transmission is fast, since it is independent of the number of bits held by the register. In terms of hardware, however, parallel transmission is less economical. A separate flip-flop and a separate transmission link is required for each bit of data. Also, in integrated circuit form the registers are limited in their capacity by the available number of pins.

Many parallel registers have serial input and serial output also available.

A simple way to load a register synchronously is to use a row of single pole double throw (SPDT) switches connected as shown in Fig. 8.14. With the clock input at ground level, logic '0', the switches are set to the required word, say 1011. The data input from the switches is not recognised by the flip-flops until the clock pulse is received. Assuming TTL bistables, the output can be monitored by connecting a light emitting diode (LED) to ground from each output via a current limiting resistor. By changing the switch settings and pulsing the clock line, simultaneous transfer of data to the outputs can be observed.

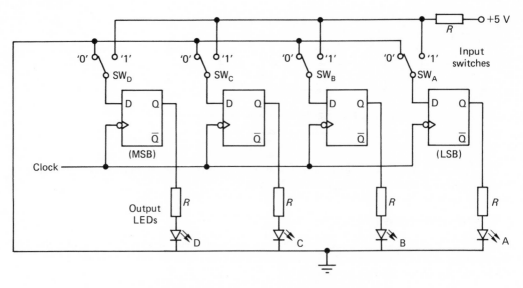

Fig 8.14   Shift register with parallel data transfer

Example 8.5
Parallel input shift register
A parallel input shift register is shown in Fig. 8.15. The circuit uses D-type flip-flops and NAND gates for entering data, i.e. **writing**, to the register.

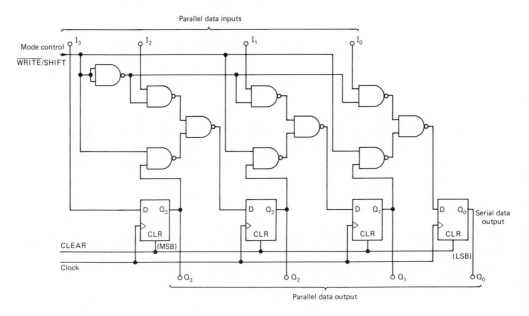

Fig 8.15   PIPO shift register

To write data in, the mode control line is taken to a logic 0, i.e. $\overline{\text{WRITE}}$ is active low and the data is clocked in. The data can be shifted when the mode control line is at a logic 1, i.e. SHIFT is active high. The register performs right shift operation on the application of a clock pulse.

Example 8.6
Another version of PIPO with serial input and serial output can be constructed using JK flip-flops, as shown in Fig. 8.16.

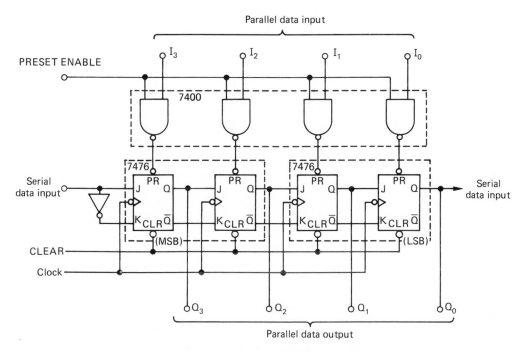

Fig 8.16   Shift register with serial I/O facility

Here, the most significant bit stage, MSB, which is also the serial input stage, is converted to a D-type flip-flop by connecting J to K via an inverter. The remaining stages have direct connections from Q to J and $\overline{Q}$ to K as shown. To enter parallel data, the register must first be cleared by the application of a logic 0 to the clear line. When the clear line is returned to logic 1, parallel data can be entered by setting the PRESET ENABLE line to logic 1. The PRESET ENABLE line must be returned to 0 before any serial operations can be carried out.

The data held in the register can be transferred to another register in the parallel mode. This is accomplished by connecting the parallel outputs Q_0, Q_1, Q_2 and Q_3 to the parallel inputs of the destination register. Alternatively, the contents of the register can be transferred serially by taking the output from the serial data output terminal and applying four clock pulses.

Example 8.7
IC version of parallel access shift register
A useful example of a parallel access shift register is the TTL 7495 integrated circuit version. This register features both parallel and serial inputs, parallel outputs, shift right

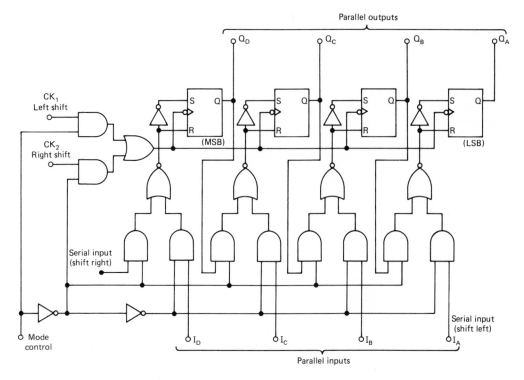

Fig 8.17   Functional diagram of 7495 4-bit shift register

and shift left operations, for which there are two separate clock inputs. A functional diagram of the four-bit register is shown in Fig. 8.17.

The clock input may be applied commonly if both CK_1 and CK_2 can be clocked from the same source. Changes at the mode control input should be made while both clock inputs are low.

For parallel loading, the data is loaded into the associated flip-flops and appears at the outputs after the high-to-low transition of the CK_2 input.

For serial loading, the mode control should be low, and data appears at the output of the flip-flop after a high-to-low transition of the CK_1 input (right shift). After four clock pulses, the least significant bit of the input data appears at the Q_A output and the most significant bit appears at Q_D.

For right shift, the serial input is held low and after the high-to-low transition of the CK_1 input the data in all flip-flops is transferred one stage to the right.

A left shift requires three connections to be made. The Q_A output must be connected to I_B input, the Q_B output to I_C input, and Q_C output to I_D input. The mode control is taken high and I_A input low. Alternatively, the data can be entered through I_A, which then acts as the serial input for the left shift data.

Example 8.8
Data bus interfacing
The circuit shown in Fig. 8.18 is an interface between a register and a data bus which permits data to be read from the register to the data bus and also allows data to be written into the register from the data bus.

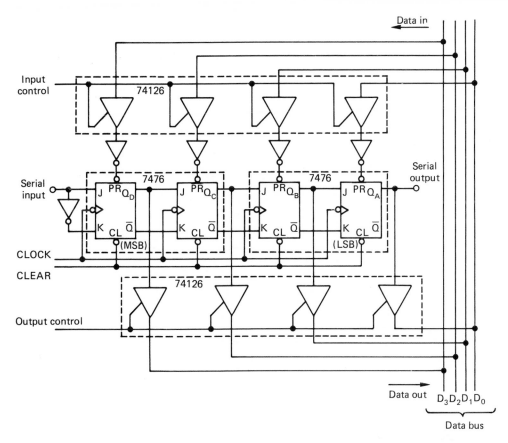

Fig 8.18 Data bus interfacing

It uses only five integrated circuits: two dual JK flip-flops (7476), two bus buffer gates with tri-state outputs (74126), and additional single inverter to convert the first stage of the register to a D-type flip-flop.

The operation is as follows. When both the input control line and the output control line are low, the register is isolated from the data bus. The register can then be serially loaded by an application of a data word to the serial input and applying four clock pulses. If the control output line is taken high, then the word stored in the register is transferred to the data bus.

If the control input line is taken high, then the data bus content is applied via inverters to the PRESET inputs of the four flip-flops. If the register is cleared prior to this data transfer, the contents of the data bus will be written into the register.

Example 8.9

The register in Example 8.8 can alternatively be implemented by the 74295 IC. It is a four-bit, bidirectional, universal shift register with tri-state outputs. A functional diagram of the register is shown in Fig. 8.19.

Here, inhibiting the entry of serial data and taking the mode control input high allows parallel loading of data into the associated flip-flops. The data appears at the outputs of

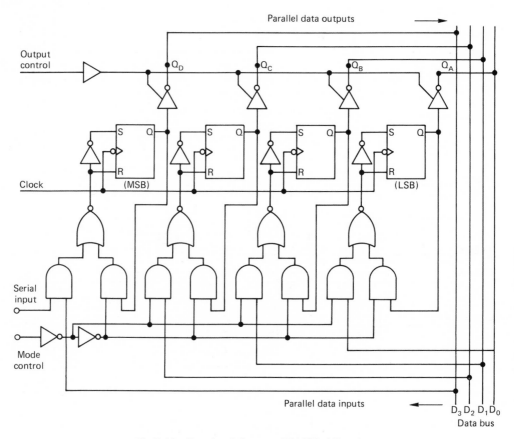

Fig 8.19 Functional diagram of 74295 shift register

the flip-flop after the high-to-low transition of the clock pulse. When the output control line is set low, the outputs are disabled and present a high impedance to the bus lines.

Sequential operation of the registers is not affected by the output control so that data can be manipulated, i.e. shifted right, and new data can be entered serially by taking the mode control low and applying the new data via the serial input.

The data can be then transferred to the bus lines by taking the output control high. The register thus acts as a serial-to-parallel converter, or as a parallel-to-serial converter. The register can be easily expanded to N bits, if required, but there are other useful registers in the integrated circuit form which can handle more than four bits.

The great variety and popularity of registers is highlighted by the number of types now available in the integrated circuit, dual-in-line, form, in the TTL 7400, CMOS 4000 and in the new high speed CMOS 74HC family range. Some of these registers are listed in Tables 8.2, 8.3 and 8.4. These are by no means complete; many other types are currently offered by various manufacturers.

Table 8.2 TTL shift registers

Device type number	Description
7491	8-bit, SISO, gated input, SR
7494	4-bit, SISO, PISO, SR
7495	4-bit, PIPO, Serial input, SR
7496	5-bit, SIPO, PISO, SR
74164	8-bit, SIPO, SR
74165	8-bit, S/PISO, SR, complementary outputs
74166	8-bit, S/PISO, SR
74194	4-bit, bidirectional universal, SR
74195	4-bit, PIPO, SIPO, PISO
74198	8-bit, bidirectional universal
74199	8-bit, bidirectional, $J\overline{K}$ serial inputs
74299	8-bit, universal shift/storage register, 3-state outputs
74323	8-bit, universal shift/storage register, 3-state outputs

Table 8.3 CMOS shift registers

Device type number	Description
4006	18-stage static shift register
4014	8-bit, SISO, PISO static shift register
4015	Dual 4-bit, static shift register
4021	8-bit, static shift register
4031	64-stage, static shift register
4035	4-bit, universal shift register
4062	200-stage, SISO, dynamic shift register
4094	8-bit, shift and store register
40100	32-stage SISO, static shift register
40194	4-bit, bidirectional universal shift register
4517	Dual 4-bit, static shift register
4557	1 to 64-bit variable length shift register
4731	Quadruple 64-bit, static shift register

Table 8.4 High speed CMOS shift registers, 74HC family

Device type number	Description
74HC164	8-bit, SIPO, shift register
74HC165	8-bit, PISO, shift register
74HC166	8-bit, S/PISO, shift register
74HC173	4-bit, D-type shift register, with 3-state outputs
74HC194	4-bit, bidirectional universal shift register
74HC195	4-bit, parallel access, shift register
74HC229	8-bit, universal shift register
74HC649	Octal bus transceiver, shift register
74HC40104	4-bit, universal shift register, with 3-state outputs
74HC4015	Dual 4-bit, SIPO, shift register
74HC4094	8-bit, shift-and-store bus, shift register

8.8 MOS and CMOS shift registers

Shift registers in the 4000 series of integrated circuits provide a comprehensive collection of circuits using MOS technology. A large number of these registers are equivalent in their operation to the 7400 series, the TTL bipolar devices.

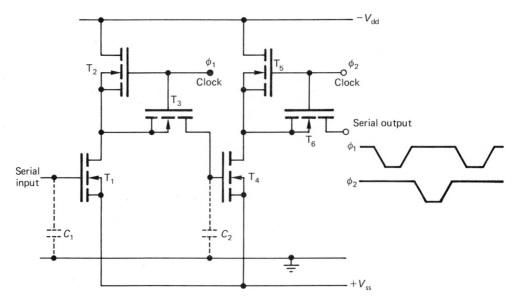

Fig 8.20   Dynamic MOS shift register cell

CMOS technology offers a much higher packing density, and registers available in this family are capable of extending the shift register wordlength to higher values than these possible with bipolar devices.

CMOS registers are commonly used for storing and shifting long binary words extending to 32, 64, 512 bits or more. These registers operate in serial mode only, storing many binary words in serial formats. Typical applications are memories for various dedicated systems, continuous display systems, calculators, delay lines, etc.

There are two types of MOS shift registers: **dynamic** and **static**.

MOS shift register cell

A storage cell of a **dynamic** shift register consists of two inverters which store information on their gate capacitances, as shown in Fig. 8.20. The inverters T_1 and T_4 are coupled together by means of T_3. The devices T_2 and T_5 serve as the nominal loads, and T_6 couples the cell to the next stage. Two non-overlapping clocks are required for proper operation. In Fig. 8.20, ϕ_1 and ϕ_2 are the two non-overlapping phases of the same clock waveform. During the time when ϕ_1 is low, data stored on C_1 is transferred to C_2 through T_3 and during the time when ϕ_2 is low the data from C_2 is transferred to the next stage of the register.

In order to retain the information the data must be recycled, i.e. the charge stored on C_1 or C_2 must be refreshed within the minimum time specified for the device, otherwise the data will be lost.

Static MOS shift registers

Static MOS registers resemble flip-flops in that they are d.c. stable and do not rely on a charge stored on a capacitor. An individual cell of a static shift register takes more space on a chip, since it uses more components, and thus consumes more power than the equivalent dynamic cell. A basic static cell is shown in Fig. 8.21(a). The main difference between this

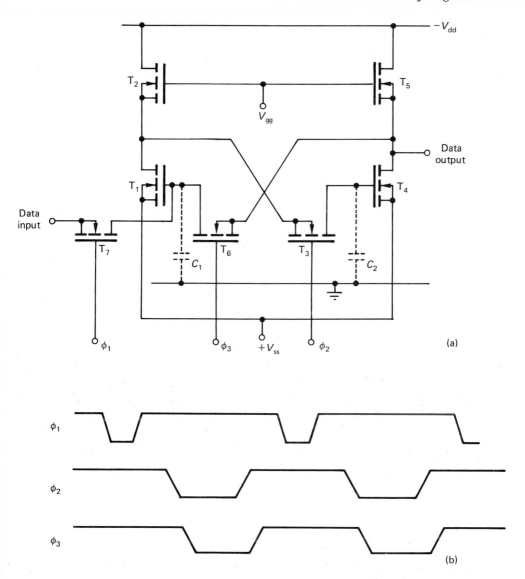

Fig 8.21   Static MOS shift register: (a) single bit cell; (b) waveforms

and the dynamic arrangement is that the loads, T_2 and T_5, are unclocked. The device T_6 now cross couples T_4 to T_1. An additional device, T_7, is used to drive the cell. The output is taken from the drain of T_4, and the ϕ_3 clock, generated internally, is identical to ϕ_2, but slightly delayed. The ϕ_3 signal is used to close the feedback loop of the flip-flop.

The respective waveforms for the stage are shown in Fig. 8.21(b).

Bipolar transistor shift registers and CMOS circuits usually operate from a single phase clock. They also use a single supply voltage rail and therefore are more popular and tend to replace other less convenient arrangements.

8.9 Problems

8.1 Explain the following terms: SISO, SIPO, PIPO and PISO.

8.2 Explain how a five-bit number can be transmitted from one register to another using serial transmission.

8.3 Explain how a recirculate facility can be added to a right shift register so that data is not lost when its contents are transmitted.

8.4 Design a six-bit serial register using JK flip-flops.

8.5 Design a four-bit PIPO register using JK bistables.

8.6 Design a five-bit SIPO register using D-type flip-flops.

8.7 Design a parallel loading circuit for a register consisting of JK master–slave bistables. Loading must be carried out via the asynchronous inputs, and should transfer the contents of a data bus directly into the register without the need to clear the previous count first.

8.8 Discuss the advantages and disadvantages of shift registers implemented in TTL and CMOS logic families.

8.9 Explain the difference between static and dynamic shift registers.

9

COUNTERS

9.1 Introduction

An electronic counter is a circuit made up of a series of flip-flops connected in a suitable manner to record sequences of pulses presented to it in digital form. Each stage of a counter will store a 0 or a 1. A counter can be designed to produce any desired sequence of bit patterns within the limits of its wordlength. The most commonly used are binary counters and counters which follow a decimal count.

As mentioned in Chapter 8, the individual flip-flops holding the least significant bit (LSB) and the most significant bit (MSB) will always be marked, whether the signal transmission followed is from left to right or from right to left. This will be illustrated below using a binary counter as an example.

Counters are basically divided into two categories:

(i) asynchronous counters, i.e. non-synchronous or ripple through counters
(ii) synchronous counters.

9.2 Asynchronous counters

Asynchronous counters, also known as ripple through or serial counters, are sequential logic systems. In these counters the first, that is the least significant, stage of the counter switches first on the application of an input pulse to the flip-flop. Successive stages change their states in turn, causing a **ripple through** effect of the count pulse. Asynchronous counters are relatively bare of additional logic elements, and so are easy to design and are relatively cheap. They are usually limited to simple counting sequences such as pure binary. Other arrangements such as BCD and up-down counters are also possible, but require additional logic gates.

The maximum count capability of an asynchronous counter depends on the number of flip-flops used. For example, a counter using four flip-flops will have a maximum count of $2^4 - 1 = 15_{10}$ or 1111_2, whereas a counter using eight stages has a maximum count of $2^8 - 1 = 255_{10} = 11111111_2$.

The total number of bit patterns that are generated by a particular counter is referred to as the **modulus** of the counter. For example, a counter, with 10 separate output patterns or

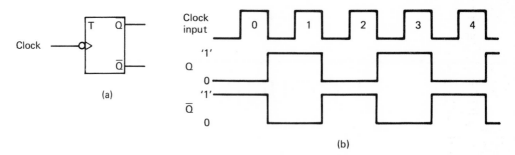

Fig 9.1   (a) Basic counter unit; (b) waveforms

counts would be a **modulo– 10** counter. In general, either SR or JK flip-flops can be used, but to operate as an asynchronous counter they must be modified to act as T-type flip-flops.

An examination of the basic, single stage, counter unit shown in Fig. 9.1 indicates that its maximum count is $2^1 - 1 = 1$. It is a **modulo– 2** counter. The output frequency is half the input frequency.

All binary counters are also frequency dividers and it is possible to design counters so that any required frequency division is obtained.

Asynchronous pure binary up counter

A four-stage ripple through binary counter is shown in Fig. 9.2(a). The four T-type flip-flops (master–slave or negative edge triggered type) are connected in cascade. The pulses to be counted are applied to the clock input of the first flip-flop. FF_A, and the output of the flip-flop drives the next flip-flop, FF_B. That is, the least significant stage of the counter is switched first and the remaining stages change state in succession, resulting in a ripple through effect of the count pulse.

The truth table for the counter. Table 9.1, shows the states of the flip-flops, and Fig. 9.2(b) shows the respective waveforms. All the flip-flops are connected to a common CLEAR line so that, before the count starts, the flip-flops can be simultaneously reset to zero.

In this counter, each successive binary stage can not change its state without the preceding stage having done so. These counters, therefore, give rise to an inherent delay in the system. The effect is cumulative, as shown in Fig. 9.2(b), and the overall delay will increase as more stages are added to the system. The rate, i.e. the highest frequency, at which the counter will reliably operate depends on the propagation delay per stage and the total number of stages used.

The counter of Fig. 9.2 is redrawn in Fig. 9.3 with the LSB at the right hand side, and with LED indicators added so that with a low frequency clock the count sequence can be followed visually.

The counter described above is universally known as an **asynchronous binary up counter** since the count is incremented by one after each input pulse. Two methods of implementing the circuit are shown in Fig. 9.4. D-type flip-flops are used in Fig. 9.4(a) and JK flip-flops are employed in Fig. 9.4(b).

Because JK flip-flops are the most popular and the most commonly available in integrated circuit form, they will be used in all future diagrams with J and K inputs connected to logic 1, to satisfy the T-type requirement. The CLEAR line is not shown, but it is assumed that both versions of the counter have this facility incorporated in order to reset the outputs to zero before commencement of a new count.

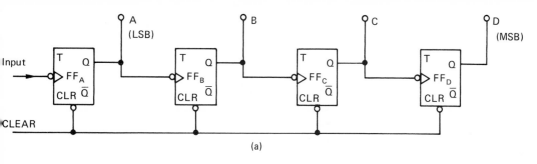

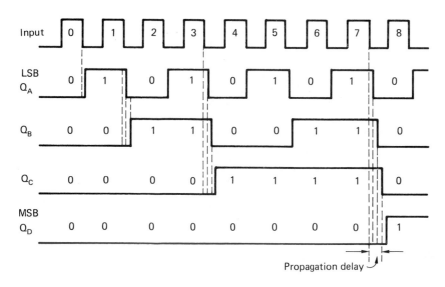

Propagation delay

(b)

Fig 9.2 (a) Ripple counter; (b) waveforms

Table 9.1 Binary ripple counter truth table

Count	D	C	B	A	Output elements			
0	0	0	0	0	$\bar{D}$	$\bar{C}$	$\bar{B}$	$\bar{A}$
1	0	0	0	1	$\bar{D}$	$\bar{C}$	$\bar{B}$	A
2	0	0	1	0	$\bar{D}$	$\bar{C}$	B	$\bar{A}$
3	0	0	1	1	$\bar{D}$	$\bar{C}$	B	A
4	0	1	0	0	$\bar{D}$	C	$\bar{B}$	$\bar{A}$
5	0	1	0	1	$\bar{D}$	C	$\bar{B}$	A
6	0	1	1	0	$\bar{D}$	C	B	$\bar{A}$
7	0	1	1	1	$\bar{D}$	C	B	A
8	1	0	0	0	D	$\bar{C}$	$\bar{B}$	$\bar{A}$
9	1	0	0	1	D	$\bar{C}$	$\bar{B}$	A
10	1	0	1	0	D	$\bar{C}$	B	$\bar{A}$
11	1	0	1	1	D	$\bar{C}$	B	A
12	1	1	0	0	D	C	$\bar{B}$	$\bar{A}$
13	1	1	0	1	D	C	$\bar{B}$	A
14	1	1	1	0	D	C	B	$\bar{A}$
15	1	1	1	1	D	C	B	A

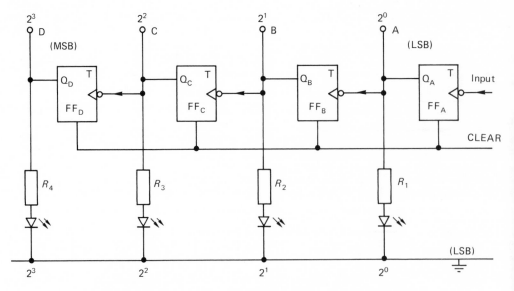

Fig 9.3 Ripple counter connections with LED output state indicators

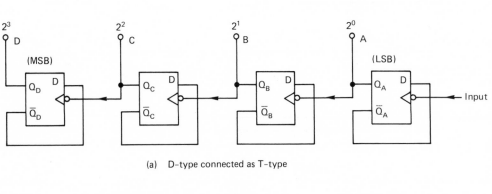

(a)   D-type connected as T-type

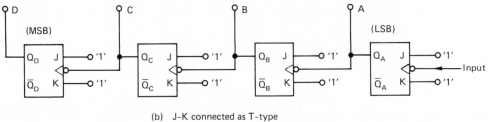

(b)   J-K connected as T-type

Fig 9.4 Two versions of the ripple through binary up counter

JK flip-flops have an added advantage over other flip-flops, in the both inputs can be connected together to a common line which then can be taken either to a logic 0 to inhibit the count or to a 1 to start counting.

Asynchronous binary down counter

An **asynchronous binary down** counter decreases its stored count by one digit (i.e. is decremented) for each input pulse.

The binary up counters described above can be modified to operate as down counters in one of the two possible ways; (a) the $\overline{Q}$ output of each stage can be used as the input to the next more significant stage, or (b) the $\overline{Q}$ outputs could be monitored instead of the Q outputs, as shown in Fig. 9.5(a) and (b).

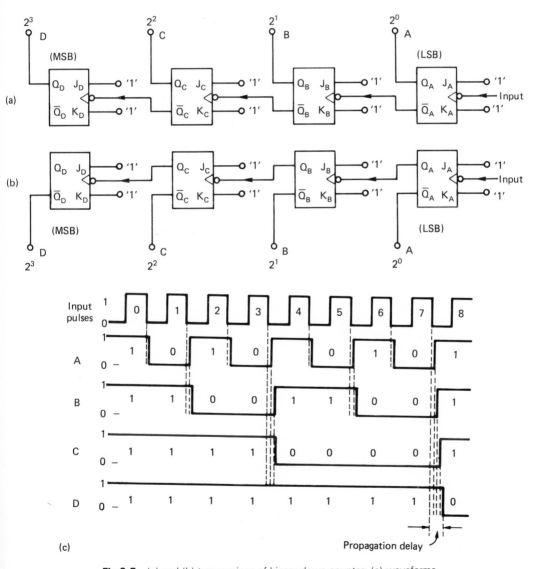

Fig 9.5 (a) and (b) two versions of binary down counter; (c) waveforms

All J and K input terminals are fed with logic 1, and therefore all flip-flops operate as T-types, switching on the high-to-low transition of the input pulse. The input pulses to be counted are applied to the clock input of the flip-flop indicating the least significant bit. Each successive stage is clocked by the output from the preceding stage. The waveforms for the counter are shown in Fig. 9.5(c), where it is clearly indicated that the counter suffers from propagation delay through the stages in the same way as the binary up counter. The truth table, Table 9.2, shows that the counter is preset to its maximum count of 1111_2 before the start of the count down.

Table 9.2   Binary down counter truth table

Input pulse sequence	D	C	B	A	Output elements			
0	1	1	1	1	D	C	B	A
1	1	1	1	0	D	C	B	$\bar{A}$
2	1	1	0	1	D	C	$\bar{B}$	A
3	1	1	0	0	D	C	$\bar{B}$	$\bar{A}$
4	1	0	1	1	D	$\bar{C}$	B	A
5	1	0	1	0	D	$\bar{C}$	B	$\bar{A}$
6	1	0	0	1	D	$\bar{C}$	$\bar{B}$	A
7	1	0	0	0	D	$\bar{C}$	$\bar{B}$	$\bar{A}$
8	0	1	1	1	$\bar{D}$	C	B	A
9	0	1	1	0	$\bar{D}$	C	B	$\bar{A}$
10	0	1	0	1	$\bar{D}$	C	$\bar{B}$	A
11	0	1	0	0	$\bar{D}$	C	$\bar{B}$	$\bar{A}$
12	0	0	1	1	$\bar{D}$	$\bar{C}$	B	A
13	0	0	1	0	$\bar{D}$	$\bar{C}$	B	$\bar{A}$
14	0	0	0	1	$\bar{D}$	$\bar{C}$	$\bar{B}$	A
15	0	0	0	0	$\bar{D}$	$\bar{C}$	$\bar{B}$	$\bar{A}$

If the J and K inputs are connected to a common control line, it can act as the count inhibit or load control, for presetting the counter, or as a stop/start control.

Asynchronous binary up/down counter

The operation of the binary up counter can be combined with that of the binary down counter to form an **up/down counter**. The new circuit is often referred to as a **reversible counter**, and is shown in Fig. 9.6.

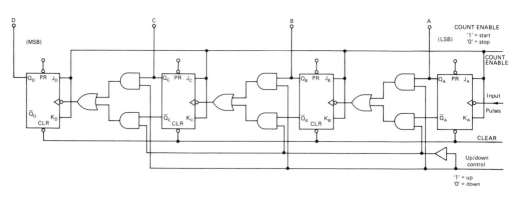

Fig 9.6   Asynchronous binary up/down counter

Here three logic gates per stage are required to switch the individual stages from counting up to counting down. The four flip-flops are used with J and K inputs connected to a common stop/start line. With the line at a logic zero, the count is inhibited; the counter can either be reset to zero, for an up count, or preset to the required value for a down count. A logic 1 applied to the count control line starts the count.

A second up/down control line is used to count up or down. Complementary signals are applied to the two AND gates so that a 1 will start a count up and a 0 will cause a count down.

Only four stages are shown, but more flip-flops can be added if required to extend the range of the counter. It must be remembered, however, that the inter-stage coupling gates introduce additional propagation delays which will limit the maximum rate at which the counter will operate.

Resettable asynchronous counters

Often it is required to limit the count to less than 2^N, where N is the number of flip-flops used. By producing an appropriate count modulus it is possible to use this counter for freuency division by a number that is not a power of 2. There are several different ways in which this can be achieved. One method is illustrated in the following example.

Suppose the count of the pure binary counter of Fig. 9.4 is to be limited to ten. All that is required is to reset the counter after the count of nine. This is done by connecting a NAND gate to the clearline, as shown in Fig. 9.7.

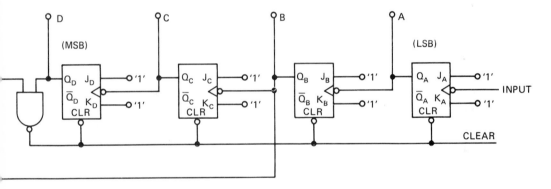

Fig 9.7 BCD 8421 decade counter

The circuit shown detects when the number $10_{10} = 1010_2$ is beginning to appear at the counter output. In this case this is represented by $Q_D = Q_B = 1$. Only when this combination appears at the input of the NAND gate will the counter reset. There is unfortunately a delay between 10_{10} appearing at the counter output and the counter resetting, which limits the operation of this counter to low frequency only.

Any modulus of count within the limits of the counter wordlength can be generated by this method. By selecting appropriate inputs to the NAND gates, any count value can be detected. The NAND gate output can be fed to appropriate PRESET and CLEAR inputs to reset the counter to any desired value. Once reset, the counter will continue as before to follow the count sequence.

In the example given of the modulo-10 counter, a resetting pulse appears once in every ten clock pulses. This can be used as a basis for frequency division, because for a Modulo-N counter a resetting will appear once every N clock pulses.

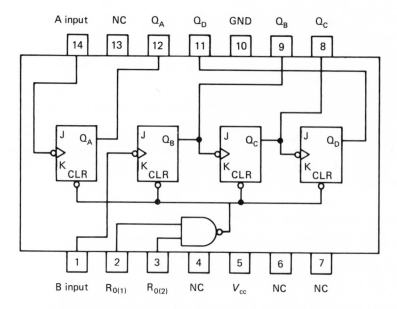

Fig 9.8   The 7493 asynchronous four-bit binary counter

An example of an asynchronous counter is the 7493, which is a four-bit binary counter with a gated RESET line which returns the four flip-flop outputs to zero. The counter, shown in Fig. 9.8, can also be used as a divide-by-two counter and divide-by-eight counter since the flip-flop A is independent of the other three flip-flops except for the common CLEAR line.

The truth tables for the four modes are shown in Table 9.3.

When used as a divide-by-two counter, only flip-flop A is connected. When used as divide-by-8 ripple through counter, the flip-flops B, C and D are employed, and the input

Table 9.3   Truth table for 7493 asynchronous binary number

Input pulse sequence	Mode 1 Divide-by-2		Mode 2 Divide-by-8			Mode 3 Divide-by-10				Mode 4 Divide-by-16			
	A		D	C	B	D	C	B	A	D	C	B	A
0	0		0	0	0	0	0	0	0	0	0	0	0
1	1		0	0	1	0	0	0	1	0	0	0	1
2			0	1	0	0	0	1	0	0	0	1	0
3			0	1	1	0	0	1	1	0	0	1	1
4			1	0	0	0	1	0	0	0	1	0	0
5			1	0	1	0	1	0	1	0	1	0	1
6			1	1	0	0	1	1	0	0	1	1	0
7			1	1	1	0	1	1	1	0	1	1	1
8						1	0	0	0	1	0	0	0
9						1	0	0	1	1	0	0	1
10										1	0	1	0
11										1	0	1	1
12										1	1	0	0
13										1	1	0	1
14										1	1	1	0
15										1	1	1	1

pulses then are applied to the flip-flop B input. It should be noted that simultaneous frequency divisions by 2, 4 and 8 are available if the output signals are taken from Q_B, Q_C and Q_D respectively.

When used as a four-bit binary ripple through counter, the output of Q_A must be connected externally to the input of the flip-flop B. The input pulses are then applied to the input of flip-flop A. The counter can be reset to zero after the count of ten by connecting B (pin 9) to pin 3, and D (pin 11) to pin 2, thus operating as a decade counter. If the count is allowed to continue to its full length, an automatic reset to zero follows after divide-by-16 step.

Another example is shown in Fig. 9.9. Here only three flip-flops are used. A part of the 7493 binary counter is connected as a divide-by-six counter. On the application of the sixth input pulse the counter is reset to zero as shown in the truth table of Table 9.4.

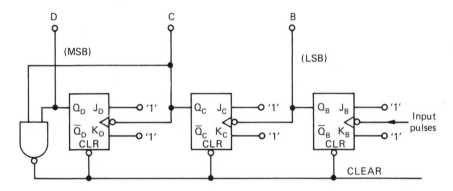

Fig 9.9 A part of the 7493 binary counter (divide by six)

Table 9.4 Divide-by-six truth table

Input pulse sequence	D	C	B
0	0	0	0
1	0	0	1
2	0	1	0
3	0	1	1
4	1	0	0
5	1	0	1
6	1	1	0
	0	0	0

7490 decade counter

A useful integrated circuit counter is the 7490 ripple through counter. Basically it provides a divide-by-two counter and a divide-by-five counter, but with its internal gating arrangements the 7490 can be connected to count in a number of different ways. Several possible connections will be shown first, leading to some useful circuits. These will indicate how to connect a number of 7490 units together to form a digital system. Figure 9.10 shows a functional block diagram of internal connections, and Table 9.5 shows the truth tables for the connections considered here.

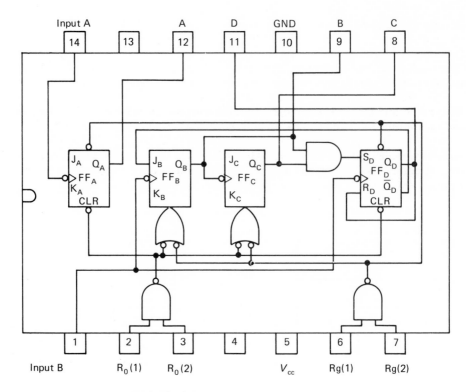

Fig 9.10   Functional diagram of 7490 counter

Table 9.5   Truth tables for 7490 connections

Input pulse	Divide-by-2 A	Divide-by-3 C	Divide-by-3 B	Divide-by-5 D	Divide-by-5 C	Divide-by-5 B	Divide-by-6 C	Divide-by-6 B	Divide-by-6 A	Decade counter D	Decade counter C	Decade counter B	Decade counter A	Divide-by-10 symmetrical A	Divide-by-10 symmetrical D	Divide-by-10 symmetrical C	Divide-by-10 symmetrical B
0	0	0	0	0	0	0	0	0	0	0	0	0	0	0	0	0	0
1	1	0	1	0	0	1	0	0	1	0	0	0	1	0	0	0	1
2	0	1	0	0	1	0	0	1	0	0	0	1	0	0	0	1	0
3		1	1	0	1	1	0	1	1	0	0	1	1	0	0	1	1
4		0	0	1	0	0	1	0	0	0	1	0	0	0	1	0	0
5				0	0	0	1	0	1	0	1	0	1	1	0	0	0
6							0	0	0	0	1	1	0	1	0	0	1
7										0	1	1	1	1	0	1	0
8										1	0	0	0	1	0	1	1
9										1	0	0	1	1	1	0	0
10										0	0	0	0	0	0	0	0

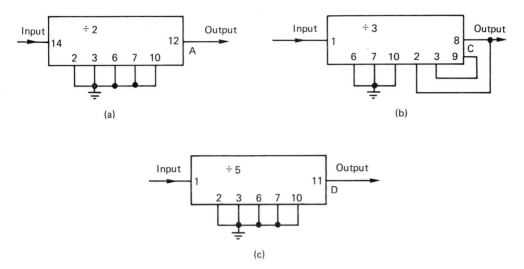

Fig 9.11 Divide-by-2, divide-by-3 and divide-by-5 connections

The counter will now be shown in block diagramatic form, highlighting the various implementations, with the necessary external connections. When used as a divide-by-two, only the flip-flop A, FF_A is employed. The input is applied to pin 14 and the output is taken from A, at pin 12. Pins 2, 3, 6, 7 and 10 are connected to ground, as shown in Fig. 9.11(a).

When used as a divide-by-three counter, the flip-flops FF_B and FF_C are used. The FF_B holds the least significant bit and FF_C indicates the most significant bit. FF_A and FF_D are not used at all in this case. The input is applied to pin 1. The counter is reset to zero by returning B and C to pins 3 and 2 respectively. The output is taken from C, at pin 8, as shown in Fig. 9.11(b).

The third connection, shown in Fig. 9.11(c), is that of the divide-by-five counter. Again the flip-flop FF_A is left unconnected. FF_B indicates the LSB, and FF_D shows the MSB. The input is connected to pin 1 and the output is taken from D at pin 11. Pins 2, 3, 6, 7 and 10 are returned to ground. Due to the internal connections the counter is automatically reset to zero, as indicated in Table 9.5.

The 7490 can operate as two separate counters. This is possible when it is connected as the divide-by-two and divide-by-three, or when it is used as two independent counters divide-by-two and divide-by-five, but in either case both counters are reset simultaneously which may not be allowed to happen in most cases.

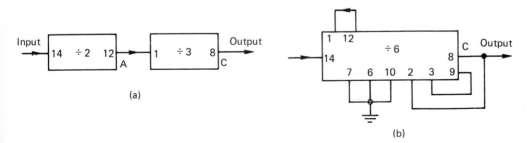

Fig 9.12 Divide-by-six counter

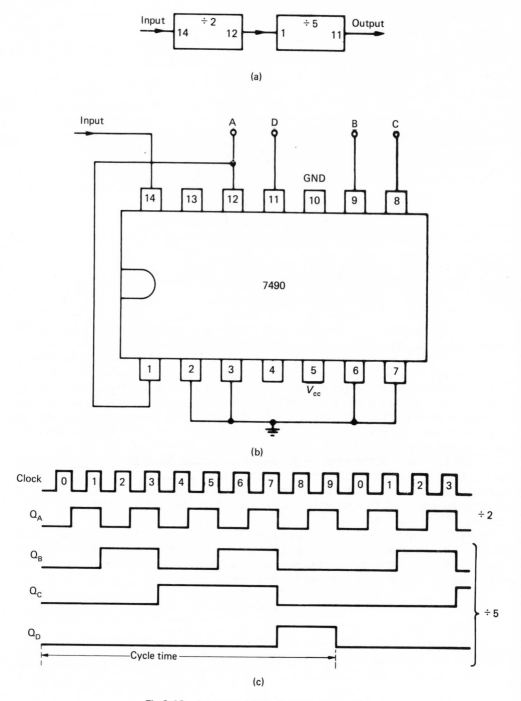

Fig 9.13　Asymetrical BCD divide-by-ten counter

For operation as a divide-by-six counter, the flip-flops FF_A, FF_B and FF_C are used. FF_A is used to indicate the LSB and FF_C to indicate the MSB of the count. An external connection from pin 12 to pin 1 is required. This connection is a combination of a divide-by-two and a divide-by-three circuit, as shown in Fig. 9.12.

The 7490 can operate in two different modes as a decade counter. When used as a binary coded decimal (BCD) counter, the input signal is applied to the divide-by-two counter and then the signal is fed to the divide-by-five counter as shown in Fig. 9.13. As shown in the truth table (Table 9.5), the output at A is the LSB and the output at D is the MSB. The output at D is that of the divide-by-ten, but the waveform is not symmetrical. Pins 2, 3, 6, 7 and 10 are grounded. The connections are shown in Fig. 9.13(b) and the output waveform in Fig. 9.13(c).

It is worth mentioning that, in addition to zero reset, the counter is provided with preset to BCD 9 count for nine's complement decimal applications (see Chapter 12).

For operation in a divide-by-ten, symmetrical mode, i.e. 1:1 square wave output, the necessary connections are shown in Fig. 9.14(a) and (b). Note that when using this connection, the sequence of the count is not sequential but is as shown in the truth table 9.5 under divide-by-10 symmetrical column, where B is the LSB and A is the MSB. The order of the count is irrelevant, however, since in frequency divisions only one output pulse is provided for every ten input pulses. The output in this case is taken from A to pin 12. The waveforms for the counter are shown in Fig. 9.14(c).

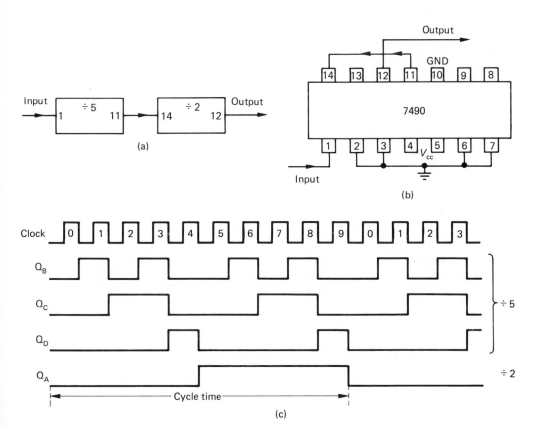

Fig 9.14 Symetrical divide-by-ten counter

9.3 Synchronous counters

In synchronous counters, flip-flops are operated by a common clock pulse and thus change their state simultaneously. In order to achieve the desired count, a suitable logic circuit is needed to determine the required state of each stage after an application of a clock pulse and set the value of the synchronous inputs accordingly. In a synchronous counter the propagation delay is that of a single stage, and therefore it will operate at much higher frequency than an equivalent asynchronous counter.

The design of a synchronous counter involves:

(a) writing a truth table for the count
(b) deriving a minimal logic expression for each J and K input using Karnaugh maps
(c) implementing the logic with a suitable set of gates
(d) building and testing the counter.

In general, synchronous counters are designed around JK flip-flops, because these flip-flops are readily available in the integrated circuit form. However SR and D-type flip-flops can also be employed.

Synchronous binary counters

Two-stage synchronous binary up counter
A two-stage synchronous binary up counter will now be designed using JK flip-flops.

Table 9.6 shows the output state change of the flip-flop, for given input conditions, after an application of a clock pulse. The X's mark the 'don't care' conditions, which are represented by either a 0 or a 1. This means that the desired output change can be effected either if $X = 0$ or it $X = 1$, The advantage of this is that hardware can be minimised if these don't care conditions are known.

Table 9.6   J–K output transition table

Output state change Q_n	Q_{n+1}	Inputs J	K
0	0	0	X
0	1	1	X
1	0	X	1
1	1	X	0

The output state change table is then written for the counter, as shown in Table 9.7 where A is the LSB. The corresponding Karnaugh map for the binary code is shown in Fig. 9.15.

In this map each cell is marked with its decimal equivalent, which is of great help in preparing the output state change maps of Fig. 9.16. It is much easier to refer to decimal numbers representing the cell positions, which at the same time are the input count pulses.

Table 9.7   Counter state table

Input pulse	Present state Q_n B	A	Next stage Q_{n+1} B	A
0	0	0	0	1
1	0	1	1	0
2	1	0	1	1
3	1	1	0	0

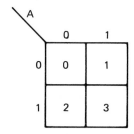

Fig 9.15 Karnaugh map for binary code sequence

The usefulness of these maps will become apparent when counters with more stages are considered later.

Six maps in all are shown in Fig. 9.16. Two maps are for the flip-flop output transition states, and the remaining four maps are for each of the flip-flops' input requirements. The information from Table 9.7 is first transferred to the output code change maps, in Fig. 9.16, for Q_A and Q_B. Then, using Table 9.6 and Fig. 9.15, the input maps for J and K are completed.

The procedure is as follows. Considering one column at a time, the values of Q_n and Q_{n+1} are transferred one-by-one from Table 9.7 to the the corresponding output state change Karnaugh map of Fig. 9.16. Thus for the input count of 0, Q_A changes from 0 to $Q_{A+1} = 1$. Therefore 0–1 is written in cell 0 of the map for Q_A in Fig. 9.16. Next, for the input count of 1, A changes from 1 to 0; therefore 1–0 is written in the corresponding cell 1. Then 0–1 is entered in cell 2, and 1–0 in cell 3.

Similarly, the values for Q_B and Q_{B+1} are transferred, one by one, from Table 9.7 to the

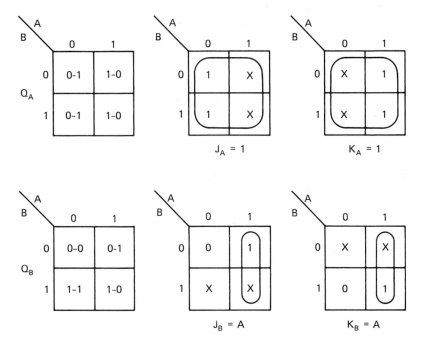

Fig 9.16 Karnaugh maps for each stage of the counter

output state change map for Q_B in Figure 9.16, in the following order: 0–0, 0–1, 1–1 and 1–0.

The next step is to complete the Karnaugh maps for the J and K inputs with the help of Table 9.6. In cell 0 of Fig. 9.16 for Q_A. the code change is 0–1, therefore 1 is written in the corresponding cell for J_A and X is entered in the corresponding cell for K_A. In cell 1, Q_A changes from 1–0, i.e. $J_A = X$ and $K_A = 1$, and so on. The transfer continues until all cells are covered.

It is now necessary to loop the 1's with as many 'don't care terms as possible, in order to extract the simplified expressions for the J and K inputs as shown in Fig. 9.16. All that remains to be done is to implement the design (Fig. 9.17). i.e. to build the circuit and test its operation.

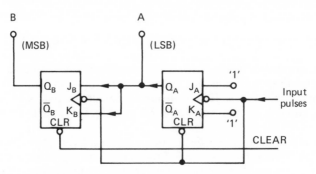

Fig 9.17 Two stage synchronous binary up counter

A CLEAR line is normally included to reset the counter initially. It is a cyclic counter, as the counter state table indicates, so that it automatically resets itself to zero on completion of the third input pulse. If the clock pulses are not interrupted, the count will continue over and over again.

Four-stage synchronous binary up counter

The design of the binary up counter is extended here to a four-stage counter. The design procedure is similar to that for the two-stage counter, except that more Karnaugh maps are needed. The Karnaugh map for binary code sequence is given in Fig. 9.18.

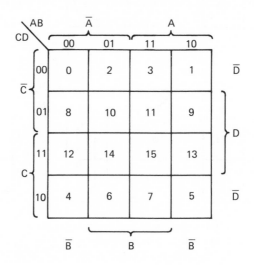

Fig 9.18 Karnaugh map for the binary code sequence

Table 9.8 Four-stage binary up counter truth table

Input pulses	Present state Q_n				Next state Q_{n+1}			
	D	C	B	A	D	C	B	A
0	0	0	0	0	0	0	0	1
1	0	0	0	1	0	0	1	0
2	0	0	1	0	0	0	1	1
3	0	0	1	1	0	1	0	0
4	0	1	0	0	0	1	0	1
5	0	1	0	1	0	1	1	0
6	0	1	1	0	0	1	1	1
7	0	1	1	1	1	0	0	0
8	1	0	0	0	1	0	0	1
9	1	0	0	1	1	0	1	0
10	1	0	1	0	1	0	1	1
11	1	0	1	1	1	1	0	0
12	1	1	0	0	1	1	0	1
13	1	1	0	1	1	1	1	0
14	1	1	1	0	1	1	1	1
15	1	1	1	1	0	0	0	0

The four-stage binary up counter truth table is shown in Table 9.8. The design continues with the preparation of the Karnaugh maps, Fig. 9.19, and the derivation of the minimal logic expressions for the J and K inputs. The counter is finally implemented as shown in Fig. 9.20.

Examining the two designs of the synchronous binary up counters, it can be seen that a pattern develops which is a useful guide for an expansion to n stages.

All that is needed is a two-input AND gate for each additional stage, since for the fifth stage $J_E = K_E = ABCD = J_D.D$; and in general $J_n = K_n = ABC \ldots (n-1)$. This means that the next added stage will not change its state until the outputs of all the preceding stages are at logic 1.

The above development clearly indicates that the logic interconnecting the stages can be implemented in two different ways. The first uses implementation of the expressions obtained from the Karnaugh maps, i.e. each J and K input is fed via a separate multi-input AND gate with all the inputs fed in parallel, as shown in Fig. 9.21(a). The second uses a serial chain with parallel entries of each successive output to form an input to the next stage, as shown in Fig. 9.21(b), thus only a two-input AND gate is needed for each additional stage. The coupling circuit is often implemented with NAND gates by replacing each AND gate with a NAND gate and an inverter, i.e. using two two-input NAND gates. The second NAND gate has the two inputs strapped together to act as an inverter, as shown in Fig. 9.21(c). The advantage of the serial chain with parallel entries implementation is that simpler gates are employed. In fact, the same input gates are used per stage irrespective of the number of stages, whereas in the direct parallel input implementation each successive flip-flop requires a gate with an additional input, i.e. non-standard gates or a combination of existing gates.

The disadvantage of the serial chain coupling is that the propagation delay increases more rapidly with each added stage, which reduces the upper frequency limit of the counter.

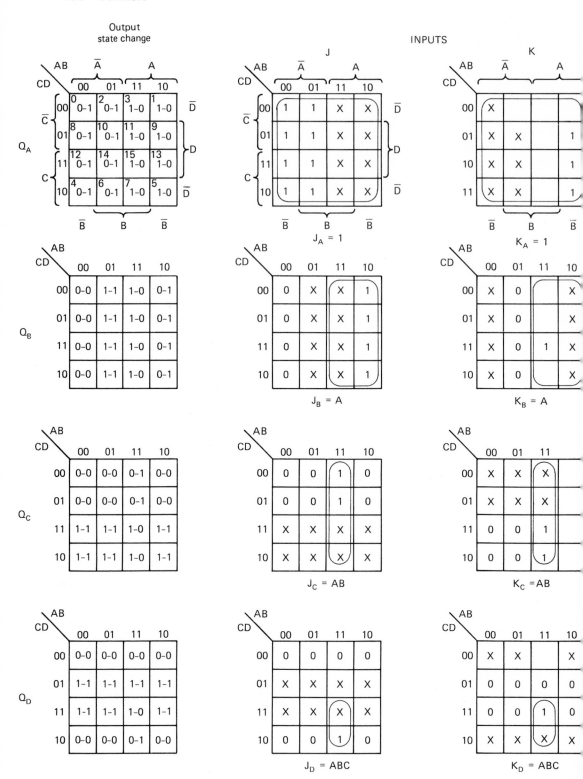

Fig 9.19 Karnaugh maps for four-stage binary up counter

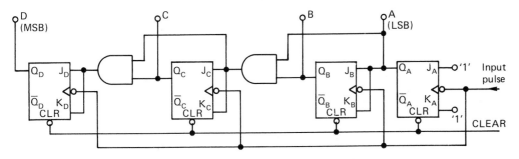

Fig 9.20 Four-stage synchronous binary up counter

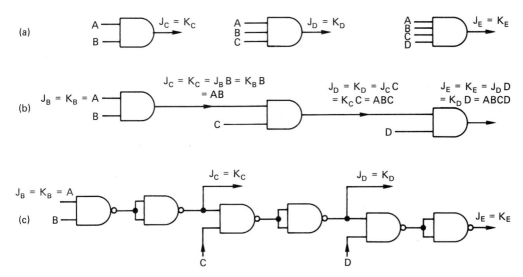

Fig 9.21 Interstage coupling logic: (a) parallel input; (b) serial chain with parallel entries using AND gates; (c) serial chain with parallel entries, using NAND gates

9.4 Synchronous decade counters

Although most industrial switching, control and computing application use pure binary code, it is often more convenient to display the results using decimal numbers. The BCD 8421 code is the most frequently used of the range of possible codes to display decimal numbers. Therefore a number of design examples will use this code. Counters using other codes will also be considered.

The BCD 8421 code uses four binary digits, but out of the sixteen possible combinations only the first ten are used. (See Chapter 12.) The remaining six numbers are treated as redundant, or don't care, states and as such are marked with X's, both in the truth tables and Karnaugh maps.

BCD 8421 up counter using S–R flip-flops

For SR flip-flops, the output state transitions are shown in Table 9.9.

Table 9.9   SR output state transition table

| Output state change | | Inputs | |
Q_n	Q_{n+1}	S	R
0	0	0	X
0	1	1	0
1	0	0	1
1	1	X	0

The counter state truth table is given in Table 9.10 and the Karnaugh map for the decimal code sequence is shown in Fig. 9.22. The variable A is the least significant bit, and X's indicate the don't care states.

Table 9.10   Truth table for BCD 8421 up counter

Input pulses	Present state Q_n				Next state Q_{n+1}			
	D 8	C 4	B 2	A 1	D 8	C 4	B 2	A 1
0	0	0	0	0	0	0	0	1
1	0	0	0	1	0	0	1	0
2	0	0	1	0	0	0	1	1
3	0	0	1	1	0	1	0	0
4	0	1	0	0	0	1	0	1
5	0	1	0	1	0	1	1	0
6	0	1	1	0	0	1	1	1
7	0	1	1	1	1	0	0	0
8	1	0	0	0	1	0	0	1
9	1	0	0	1	0	0	0	0
10 to 15	X	X	X	X	X	X	X	X

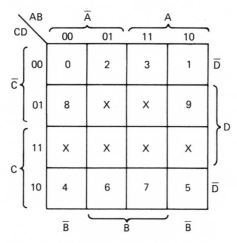

Fig 9.22   Karnaugh map showing decimal code sequence

The design proceeds with preparation of the Karnaugh maps in Fig. 9.23 from which Boolean expressions are derived for the inputs of the SR flip-flops.

From Fig. 9.23 it is apparent that there are two possible solutions for the R_D input logic. The first solution is implemented as shown in Fig. 9.24, where two two-input AND gates and three three-input AND gates are required. But, when the Boolean expressions obtained

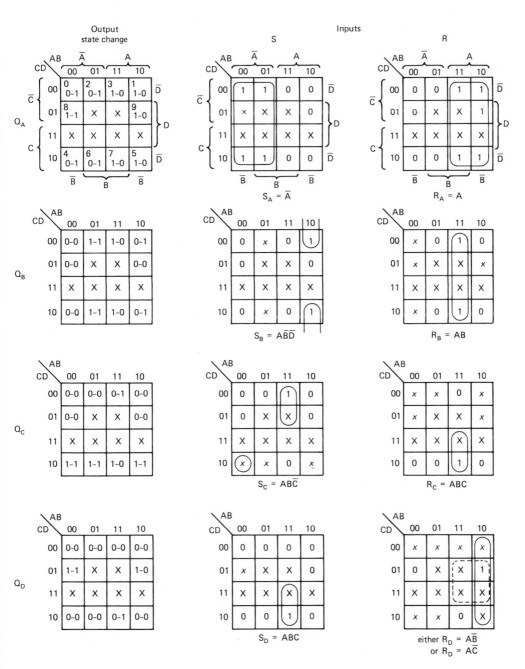

Fig 9.23   Karnaugh maps for S-R BCD 8421 up counter

from the Karnaugh maps are further manipulated, the final circuit is implemented with only five two-input AND gates as shown in Fig. 9.25. A common CLEAR line is normally provided as indicated on both diagrams.

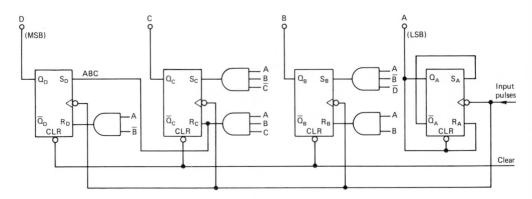

Fig 9.24 S-R BCD 8421 up counter

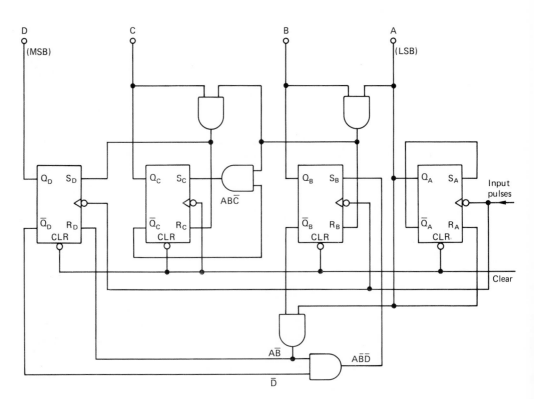

Fig 9.25 S-R BCD 8421 up counter with simplified gating system

9.5 Shift register counters

Ring counters

A natural progression from a shift register is a **ring counter**. A ring counter is a shift register with feedback from output to input. Ring counters are useful for producing timing pulses. Only one extra connection is required in order to make a counter out of a shift register: by feeding the serial output Q of the shift register back to the J input of the first stage. This assumes that an inverter is used from J to K, at the input to the shift register, to make the first stage act as a D-type flip-flop. The inverter is not required if a second connection is made. This time the $\overline{Q}$ of the output stage is fed back to the K input of the first stage. The two possible connections are shown in Fig. 9.26, using as examples three-stage ring counters.

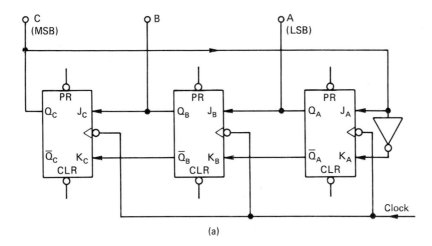

(a)

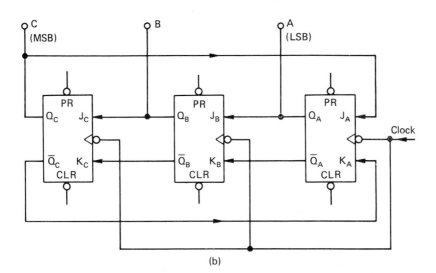

(b)

Fig 9.26 Two possible connections of three-stage ring counters

Independent CLEAR and PRESET facilities are shown for each flip-flop. An application of the clock pulses will circulate the data contained within the counter. Normal shift operation is assumed using A as the least significant bit (LSB). The connections shown in Fig. 9.26 apply to any number of stages.

Twisted ring counters

A **twisted ring counter**, or a **Johnson counter**, is also referred to as a **switch tail ring counter**. The counter is constructed from a shift register using inverse feedback. The Q of the final stage is returned to the K of the first stage, and the $\bar{Q}$ of the last stage is connected to the J input of the first stage. Thus the connections are twisted, switched or interchanged. This arrangement halves the number of stages required for a given count as compared with the ring counter.

Two-stage twisted ring counter
A two stage twisted ring counter is shown in Fig. 9.27. The circuit can be cleared or preset to 11 before counting commences, as shown in the truth tables of Table 9.11. For a two stage circuit there are $2^2 = 4$ possible output states.

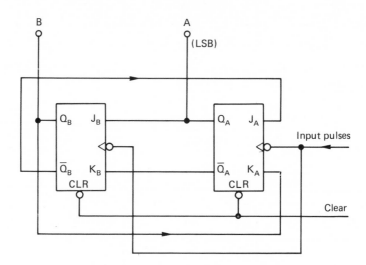

Fig 9.27   Two-stage twisted ring counter

Table 9.11   Truth tables for two-stage twisted ring counter

Output states	B	A	Input pulses	B	A	Sequence	Input pulses	B	A	Sequence
0	0	0	0	0	0	0	0	1	1	3
1	0	1	1	0	1	1	1	1	0	2
2	1	0	2	1	1	3	2	0	0	0
3	1	1	3	1	0	2	3	0	1	1
(a)			4	0	0	0	4	1	1	3
				(b)				(c)		

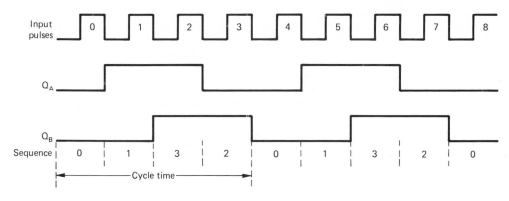

Fig 9.28 Waveforms for the two-stage twisted ring counter

When connected as a twisted ring counter, the circuit offers two possible sets of all four states as shown in Table 9.11(b) and (c). Since the order in which the decimal equivalent number of the sequence is the same in each case, there is only one unique sequence 0, 1, 3, 2. The output waveforms for the sequence are shown in Fig. 9.28.

A symmetrical, divide-by-two, waveform is obtained from either Q_A or Q_B. A CLEAR line is provided for resetting to 00 before the count commences. The resetting in this case is not really necessary since the two-stage twisted ring counter has only one sequence irrespective of the initial state of the counter.

Three-stage twisted ring counter

If a third stage is added to the twisted ring counter described above, the count is extended to six as shown by the truth tables of Table 9.12. The circuit of the three-stage twisted ring counter is shown in Fig. 9.29.

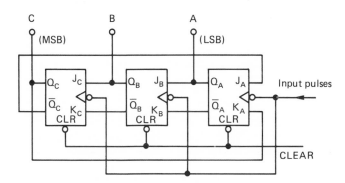

Fig 9.29 Three-stage twisted ring counter

Starting the count cycle from all clear or preset to 111 provides the same numerical order as illustrated in Table 9.12(b) and (d). Therefore the count follows one and the same sequence, 0, 1, 3, 7, 6, 4; the waveforms for this are shown in Fig. 9.30. There is another possible sequence, as shown in Table 9.12(c): and 2.5, which is never used in practice and is considered as illegal.

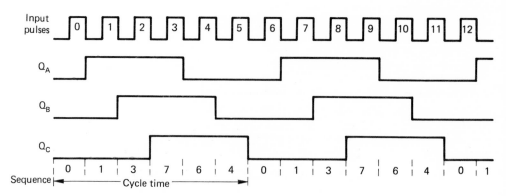

Fig 9.30　Waveforms for the three-stage twisted ring counter

Table 9.12　Truth tables for three-stage twisted ring rounter

Output states	C	B	A	Input pulses	C	B	A	Sequence
0	0	0	0	0	0	0	0	0
1	0	0	1	1	0	0	1	1
2	0	1	0	2	0	1	1	3
3	0	1	1	3	1	1	1	7
4	1	0	0	4	1	1	0	6
5	1	0	1	5	1	0	0	4
6	1	1	0					
7	1	1	1	6	0	0	0	0
	(a)				(b)			

Input pulses	C	B	A	Sequence
0	0	1	0	2
1	1	0	1	5
2	0	1	0	2
	(c)			

Input pulses	C	B	A	Sequence
0	1	1	1	7
1	1	1	0	6
2	1	0	0	4
3	0	0	0	0
4	0	0	1	1
5	0	1	1	3
6	1	1	1	7
		(d)		

As shown in Fig. 9.29 a CLEAR line is provided to reset the counter before the start of the count, so that the sequence derived in the truth table could be followed. If the output is taken from Q_C, the waveform is symmetrical, divide-by-six, as shown in Fig. 9.30.

9.6　Synchronous modulo counters

A counter can be designed to have a particular modulus or scale, i.e. it can be made to divide the frequency of the incoming input pulses by a predetermined value. The modulus of a counter is the number of possible 0 and 1 states, which for n flip-flops can never be greater than 2^n; but the actual count can be stopped earlier. The difference between the circuit operating as a counter or as a scaler is that the former has the outputs of all the flip-flops available for direct counting in binary or for decoding other sequences to a predetermined number system; the latter has only one output, usually taken from the last flip-flop in the chain. In other words a **scaler**, or a **modulus counter**, is a **frequency divider circuit**. Some counter circuits are designed such that they are capable of performing both operations. It is not necessary to use inter-stage logic in order to design modulo counters. The necessary feedback can be achieved by suitable connections of the flip-flops, as illustrated by the following examples. Often there are many different possible connections for a counter, but only one of these will be shown here for each type of the counter discussed.

Scale-of-2 (modulo-2) counter

This is the simplest modulo counter possible, employing a single JK flip-flop with both J and inputs at logic 1, as shown in Fig. 9.31(a). The truth table is at (b) and the accompanied output waveforms at (c). The circuit provides a negative-going , or falling, edge on the high-to-low transition of alternate input pulses.

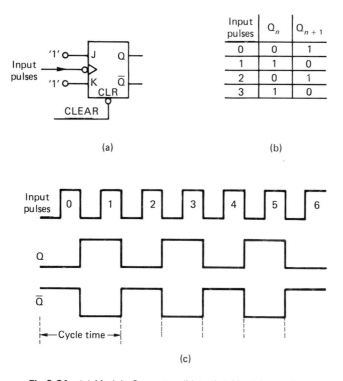

Input pulses	Q_n	Q_{n+1}
0	0	1
1	1	0
2	0	1
3	1	0

(a)   (b)

(c)

Fig 9.31   (a) Modulo-2 counter; (b) truth table; (c) waveforms

A modulo-2 counter can also be connected as a single stage twisted ring counter shown in Fig. 9.32. Here, the Q output is fed back to the K input and the $\overline{Q}$ output is connected to the J input. The truth table and the waveforms for the counter are shown in Fig. 9.32(b) and (c) respectively.

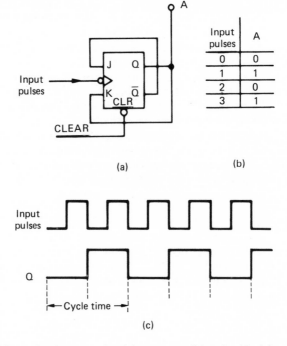

Fig 9.32 (a) Single-stage twisted ring counter; (b) truth table; (c) waveforms

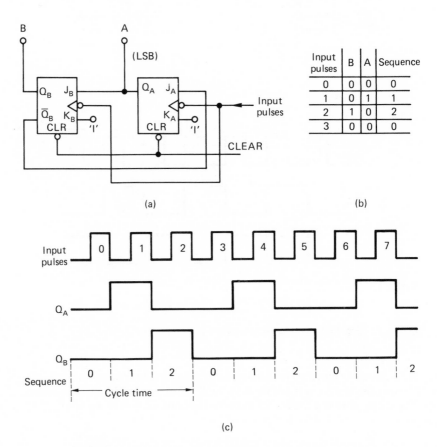

Fig 9.33 (a) Modulo-3 counter; (b) truth table; (c) output waveforms

Scale-of-3 (modulo-3) counter

Two flip-flops are required for a scale-of-3 counter. A suitable circuit is shown in Fig. 9.33, together with its truth table and the output waveforms.

The inputs to K_A and K_B are don't care states, i.e. they can either be a 0 or a 1. The $\overline{Q}_B$ output is fed back to J_A in order to hold the first stage at a 0 when the second stage is at a 1 logic level. This resets the counter to zero and the next pulse restarts the count again, as shown in the truth table in Fig. 9.33(b). The output waveforms for the counter are shown in Fig. 9.33(c), which shows that Q_B provides a negative-going edge at the end of the third pulse, i.e. on the high-to-low transition of the input pulse.

Other modulo counters

There are many other ways to design counters with a predetermined cycle length. As an example, a counter available in the integrated circuit form will be considered next. For instance, a modulo-7 counter can very easily be obtained using the 7492 integrated circuit, the divide-by-twelve counter, when the output of Q_A is connected to the input J_B.

The 7492, however, is a modulo-3 synchronous counter, formed by flip-flops FF_B and FF_C coupled asynchronously to a divide-by-two FF_D flip-flop, followed by a single divide-by-two FF_A stage.

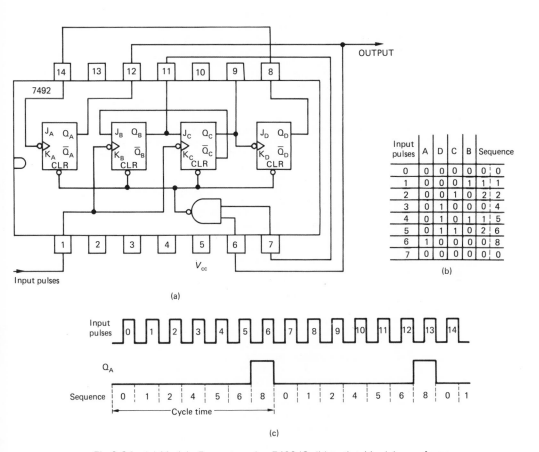

Input pulses	A	D	C	B	Sequence
0	0	0	0	0	0
1	0	0	0	1	1
2	0	0	1	0	2
3	0	1	0	0	4
4	0	1	0	1	5
5	0	1	1	0	6
6	1	0	0	0	8
7	0	0	0	0	0

(b)

Fig 9.34 (a) Modulo-7 counter using 7492 IC; (b) truth table; (c) waveforms

Table 9.13   TTL counters, 7400 series

7490	Decade, ripple, (divide-by-2 and divide-by-5)
92	Divide-by-12 (divide-by-2 and divide-by-6)
93	4-bit binary (divide-by-2 and divide-by-8)
160	4-bit synchronous, decade, direct clear
161	4-bit synchronous, binary, direct clear
162	4-bit synchronous, decade, synchronous clear
163	4-bit synchronous, binary, synchronous clear
168	4-bit synchronous up/down, decade
169	4-bit synchronous up/down, binary
176	Presetable counter/latch, decade
177	·Presetable counter/latch, binary
190	Synchronous up/down, BCD
191	Synchronous up/down, binary
192	Synchronous up/down, BCD with clear
193	Synchronous up/down, binary with clear
196	Preset table counters/latches, decade/bi-quinary
197	Preset table counters/latches, binary
290	Decade (divide-by-2 and divide-by-5)
293	4-bit binary (divide-by-2 and divide-by-8)
390	Dual decade, bi-quinary or BCD sequences
393	4-bit binary
668	Synchronous, up/down, decade
669	Synchronous, up/down, binary
690	Synchronous with register, decade
691	Synchronous with register, binary
692	Fully synchronous with register, decade
693	Fully synchronous with register, binary
696	Synchronous up/down with register, decade
697	Synchronous up/down with register, binary
698	Fully synchronous up/down with register, decade
699	Fully synchronous up/down with register, binary

Table 9.14   CMOS 4000B series, counters

4017	5-stage Johnson counter
4018	Presetable, divide-by-n counter
4020	14-bit binary counter
4022	4-stage (octal), divide-by-8 Johnson counter
4024	7-stage binary counter
4029	Synchronous up/down, binary/decade counter
4040	12-bit binary counter
4045	21-stage binary counter
4059	Programmable divide-by-n counter
4060	14-bit binary counter
4161	4-bit binary, synchronous with clear
4162	Decade, synchronous with clear
4163	4-bit binary, synchronous with clear
4510	BCD up/down
4516	Binary up/down
4518	Dual BCD up counter
4520	Dual binary up/down counter
4521	24-stage frequency divider
4522	4-bit BCD, programmable divide-by-n
4526	4-bit binary, programmable divide-by-n
4534	Real-time 5-decade counter
4553	3-digit BCD
4737	Quad static decade counter

By connecting the Q_D to the clock input of FF_A and taking the output Q_A, as shown in Fig. 9.34(a), the counter can be reset to zero on completion of the seventh pulse. This is indicated in the truth table of Fig. 9.34(b) and illustrated by the waveforms in Fig. 9.34(c).

Modulo counters with longer cycle lengths can be designed using additional stages together with logic gating circuits, if required, or by using two or more integrated circuit counters with some external gating.

Some of the available integrated circuit counters are given in Tables 9.13, 9.14 and 9.15. These are available in TTL, high speed CMOS and in the 4000 CMOS series. The lists are not complete, as many more various counters with their own type numbers are available from different manufacturers.

Table 9.15   High speed CMOS 74HC series counters

HC160	Synchronous BCD, decade counter, asynchronous reset
HC161	Synchronous 4-bit binary counter, asynchronous reset
HC162	Synchronous BCD, decade counter, synchronous reset
HC163	Synchronous 4-bit binary counter, synchronous reset
HC190	Decade up/down, preset table BCD
HC191	4-bit binary up/down, preset table
HC192	BCD, decade up/down, separate clocks
HC193	4-bit binary up/down, separate clocks
HC390	Dual decade ripple counter
HC393	Dual 4-bit binary ripple counter
HC4017	Johnson decade counter, 10 decoded outputs
HC4020	14-bit binary ripple counter
HC4024	7-stage binary ripple counter
HC4029	Binary or BCD preset table up/down counter
HC4040	21-stage binary ripple counter
HC4060	14-stage counter with oscillator
HC4518	Dual synchronous BCD counter
HC4520	Dual 4-bit binary counter

9.7 Problems

9.1 Explain the difference between asynchronous and synchronous counters.

9.2 What is meant by a modulo-N counter?

9.3 Design an asynchronous counter that will count from 0 to 5, jump over 6 and 7 and the continue counting from 8 to 15 before restarting at 0.

9.4 Design an asynchronous modulo-5 counter.

9.5 Design an asynchronous counter circuit to divide an input frequency by 7. Include additional circuitry to ensure that the output waveform has an equal mark to-space ratio.

9.6 Design a synchronous 8421 BCD up counter that uses JK master–slave slip-flops.

9.7 Design a synchronous 8421 BCD down counter that uses JK master–slave flip-flops.

9.8 Design a synchronous counter that will follow the BCD 2421 Aiken code.

9.10 Design a 4-stage Gray code counter.

9.11 Design a 4-stage ring counter and derive all possible sequences.

9.12 Design a 4-stage twisted ring counter deriving all possible sequences.

9.13 Design a synchronous module-4 counter using JK master–slave bistables.

*Note: see Chapter 12 for details of codes.

10

LOGIC CIRCUITS TO PERFORM ARITHMETIC

10.1 Introduction

A substantial part of combinational logic is involved with the design of circuits which perform arithmetic operations. The rules of binary addition and subtraction are discussed in Chapter 2. In Chapter 8 it is shown that binary data can be manipulated in either serial or parallel form. The arithmetic circuits for each of these two forms are different. For instance, in serial circuits, referred to as **serial adders** or **serial subtractors**, each bit of binary information is added or subtracted in sequence, i.e. one bit at a time. However, in parallel circuits, i.e. parallel adders or subtractors, all bits of binary information are added or subtracted simultaneously.

10.2 Addition of binary numbers

Half adder

The design of basic logic circuits to perform half adder operations are discussed in Chapter 4 as examples of the application of combinational logic. Here, the discussion will continue

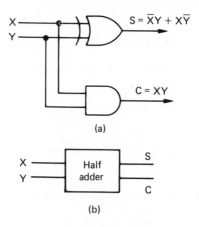

(a)

(b)

Fig 10.1 Half adder (a) X-OR; AND implementation; (b) symbol

and will be extended to other operations. Further to the circuits shown in Figs. 4.31, 4.32 and 4.33, the half adder circuit can be implemented using an exclusive-OR gate and an AND gate, as shown in Fig. 10.1(a). The output from the exclusive-OR is the sum $S = \overline{X}Y + X\overline{Y}$ and the output from the AND gate is the carry $C = XY$. Figure 10.1(b) shows the symbol of the half adder.

The circuit is called a half adder because it only operates on two binary digits and does not permit a carry input.

Full adder

In order to allow for a carry, a three-input adder circuit is required. The three inputs are X, Y and a carry from the previous operation. This can be achieved by connecting two half adders together, as shown in Fig. 10.2(a) and (b). The new circuit is called a **full adder**.

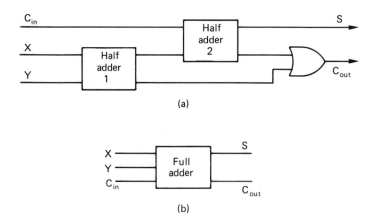

Fig 10.2 Full adder (a) implementation; (b) symbol

Table 10.1 Full adder truth table

X	Y	C_{in}	S	C_{out}
0	0	0	0	0
0	0	1	1	0
0	1	0	1	0
0	1	1	0	1
1	0	0	1	0
1	0	1	0	1
1	1	0	0	1
1	1	1	1	1

The full adder truth table is shown in Table 10.1 from which logical expressions for S and C_{out} are derived, as follows:

$$S = \overline{X}\,\overline{Y}C_{in} + \overline{X}Y\overline{C}_{in} + X\overline{Y}\,\overline{C}_{in} + XYC_{in} \text{ and}$$
$$C_{out} = \overline{X}YC_{in} + X\overline{Y}C_{in} + XYC_{in} + XYC_{in} = XY(\overline{C}_{in} + C_{in}) + (\overline{X}Y + X\overline{Y})C_{in}$$
$$= XY + (\overline{X}Y + X\overline{Y})C_{in}$$

Plotting S on Karnaugh map, Fig. 10.3(a), shows that the expression can not be simplified.

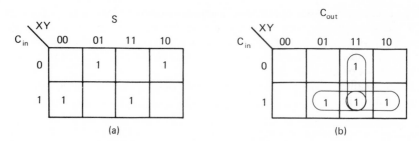

Fig 10.3   Karnaugh maps for (a) sum; (b) C_{out}

Using Boolean algebra operations, the expression can be rearranged to give

$$S = (\overline{X}\overline{Y} + XY)C_{in} + (\overline{X}Y + X\overline{Y})\overline{C}_{in} = \overline{X \oplus Y}C_{in} + (X \oplus Y)\overline{C}_{in} = X \oplus Y \oplus C_{in}$$

which can be implemented with two exclusive-OR gates, as shown in Fig. 10.4(a).

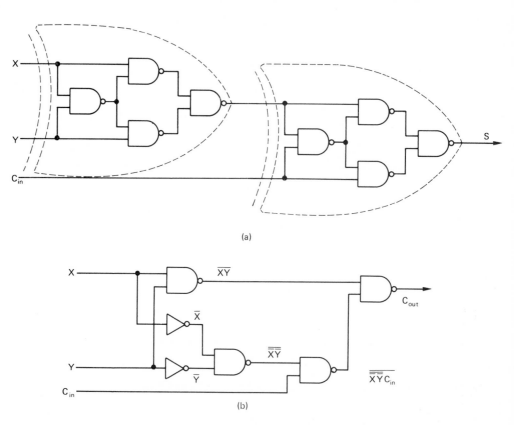

(a)

(b)

Fig 10.4   Full adder implementation: (a) sum; (b) C_{out}

Similarly, plotting C_{out} on the Karnaugh map, Fig. 10.3(b), gives

$$C_{out} = XY + XC_{in} + YC_{in}$$

Rearranging this using De Morgan's 1st theorem gives

$$C_{out} = XY + (X + Y)C_{in} = XY + \overline{\overline{X}\,\overline{Y}}\,C_{in} = \overline{\overline{XY} \cdot \overline{\overline{X}\,\overline{Y}\,C_{in}}}$$

which can be implemented with four two-input NAND gates and two NOT gates, as shown in Fig. 10.4(b). Alternatively the full adder can be implemented as shown in Fig. 10.5 by combining Fig. 10.4(a) and (b) together to form one circuit.

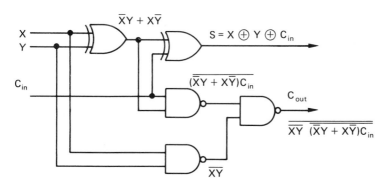

Fig 10.5 Alternate implementation of full adder using eleven NAND gates

If each XOR gate in Fig. 10.5 is implemented using four NAND gates (Fig. 10.4(a)) the circuit can further be simplified as shown in Fig. 10.6, needing only nine two-input NAND gates instead of the eleven NAND gates previously required.

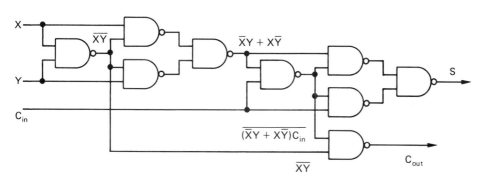

Fig 10.6 Full adder using nine NAND gates

The above circuits illustrate the principles of binary addition using logic. To perform the arithmetic operations in serial form, a few more supplementary units are needed. These are a register to hold the two numbers to be added, say X and Y; a register for the sum (an accumulator); and a single digit store to hold the carry bit. A suitable circuit is shown in Fig. 10.7. The length of each registar depends upon the length of the numbers to be added.

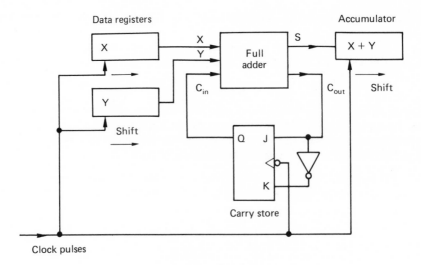

Fig 10.7 Serial adder

A separate accumulator is not really necessary. One of the two data registers could be used instead, since after addition the original numbers are shifted out and both registers are filled with 0's

A modified circuit of the serial adder is shown in Fig. 10.8. where the data register which initially holds the value of X doubles as an accumulator. Provision is also made to recirculate the data initially held in the Y register.

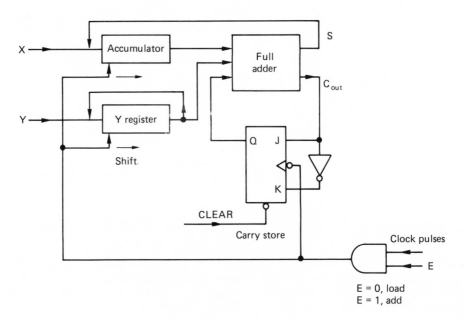

Fig 10.8 Practical serial adder circuit

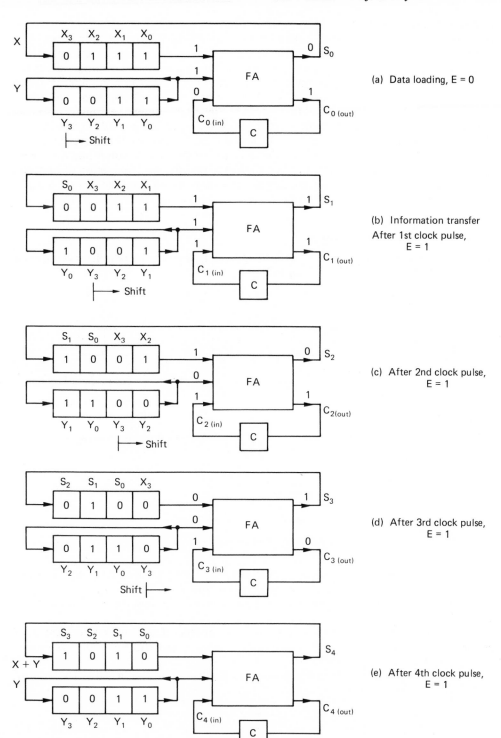

Fig 10.9 Transfer operations for a four-bit serial adder with an accumulator

A control circuit in the form of an AND gate is also incorporated in the system to allow for loading data prior to the arithmetic operations. When E = 0, the data X and Y are loaded, and the carry bit is cleared. The enable E is then taken to a logic 1 level for the duration of the number of clock pulses corresponding to the number of binary digits to be added. After the completion of the cycle the sum appears in the accumulator and the contents of the register Y are restored to the value prior to the execution of addition.

A simplified circuit of a serial adder is given in Fig. 10.9 to show the transfer operations undergone during the addition of two four-bit binary numbers, 0111 and 0011. In Fig. 10.9(a), with the control enable E = 0, X and Y are loaded into the shift registers. The control E is then taken to a logic 1 for the duration of four clock pulses and the transfer operations which follow are shown in Fig. 10.9(b) to (e) after the completion of each clock pulse. In Fig. 10.9(e) the accumulator displays the sum, X + Y, and the register Y has the original data 0011 restored in it.

10.3 Parallel full adders

Serial addition is relatively slow. In order to speed up the process parallel techniques are employed. A suitable circuit is shown in Fig. 10.10.

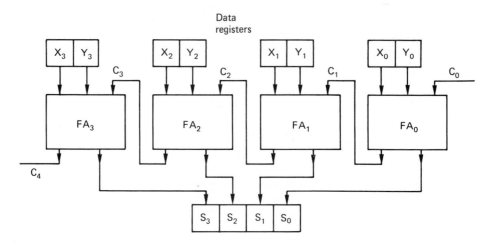

Fig 10.10   Four-bit parallel adder

The initial carry in, C_0, which is set to zero, is fed to the first full adder (FA_1). This input is provided to allow for later expansion to a larger wordlength. The carry out, C_1, from FA_1 is fed to the carry in of the second stage, FA_2, and so on. The sum from each separate full adder is fed to an accumulator to provide the four-bit sum.

The four stage counter shown in Fig. 10.10 is capable of adding two four-bit binary numbers. In order to increase the capacity of the counter an extra full adder is connected for every additional binary digit.

Full adder circuits available in integrated circuit form are listed in Table 10.2.

Table 10.2   Integrated circuit adders

7480	Single, 1-bit gated full adder
7482	Single, 2-bit full adders
7483	Single, 4-bit full adders
74283	Single, 4-bit full adders
74183	Dual, 1-bit, carry save, full adder
4008	Single, 4-bit full adder

10.4 Subtraction of binary numbers

Half subtractor

In Chapter 4 it is shown that subtraction can be performed in the same way as addition. Indeed, the same circuit can perform addition as well as subtraction as long as a borrow output is provided.

The truth table (Table 4.11) for a half subtractor is repeated here in Table 10.3 for convenience. From this, the difference $D = \overline{X}Y + X\overline{Y}$ and the borrow $B = \overline{X}Y$. These are implemented in Chapter 4, using logic gates, in Fig. 4.34 to 4.37. The implementations are further simplified here using an exclusive-OR gate, with additional gates to provide the borrow output. This is shown in Fig. 10.11(a), with the symbolic representation shown in Fig. 10.11(b).

Table 10.3   Half subtractor truth table

X	Y	D	B
0	0	0	0
0	1	1	1
1	0	1	0
1	1	0	0

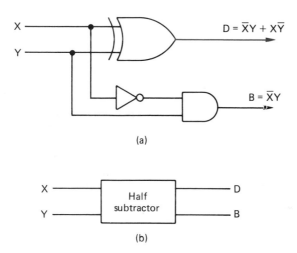

$$D = \overline{X}Y + X\overline{Y}$$

$$B = \overline{X}Y$$

(a)

(b)

Fig 10.11   Half subtractor: (a) exclusive OR, NOT, AND implementation; (b) symbol

Full subtractor

In order to provide a complete answer to subtraction, a provision for borrow must be incorporated. This is shown in Fig. 10.12(a), where two half subtractors are combined to form a full subtractor. Its symbol is shown in Fig. 10.12(b).

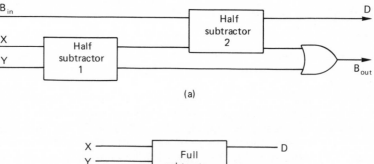

(a)

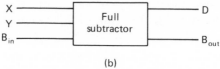

(b)

Fig 10.12　Full subtractor: (a) implementation; (b) symbol

Table 10.4　Full subtractor truth table

X	Y	B_{in}	D	B_{out}
0	0	0	0	0
0	0	1	1	1
0	1	0	1	1
0	1	1	0	1
1	0	0	1	0
1	0	1	0	0
1	1	0	0	0
1	1	1	1	1

The truth table for the three input subtractor is given in Table 10.4, from which Boolean algbra expressions for the difference and borrow are derived as follows:

$$D = \overline{X}\,\overline{Y}B_{in} + \overline{X}Y\overline{B}_{in} + X\overline{Y}\,\overline{B}_{in} + XYB_{in}$$

This is the same as the output from the full adder, where $C_{in} = B_{in}$, and therefore it can be implemented with two exclusive-OR gates in an arrangement similar to the circuit shown in Fig. 10.4(a). D can also be represented in the form:

$$D = (\overline{X}\,\overline{Y} + XY)B_{in} + (\overline{X}Y + X\overline{Y})\overline{B}_{in} = \overline{(X \oplus Y)}\,B_{in} + (X \oplus Y)\overline{B}_{in} = X \oplus Y \oplus B_{in}$$
$$B_{out} = \overline{X}\,\overline{Y}B_{in} + \overline{X}Y\overline{B}_{in} + \overline{X}YB_{in} + XYB_{in}$$

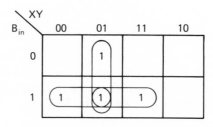

Fig 10.13　Karnaugh map for B_{out}

Using the Karnaugh map of Fig. 10.13 the expression for B_{out} can be simplified to:

$$B_{out} = XY + XB_{in} + YB_{in}$$

This expression can be implemented as shown in Fig. 10.14.

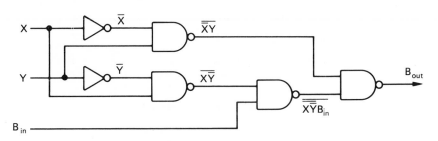

Fig.10.14   Implementation of B_{out}

The circuits for D and B_{out} are now combined using two dissimilar implementations for XOR gates. This reduces the number of NAND gates to eleven, as shown in Fig. 10.15.

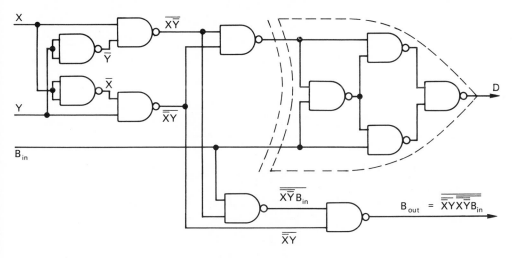

Fig 10.15   Full subtractor using eleven NAND gates

Serial and parallel subtractors

The complete serial subtractor circuit is shown in Fig. 10.16, and a parallel subtractor version is shown in Fig. 10.17.

10.5   Addition and subtraction in two's complement form

From the discussion of binary arithmetic in two's complement form in Chapter 2, and the description of adders and subtractors earlier in this chapter, it is apparent that the easiest way to perform addition and subtraction is to design a circuit to combine the two

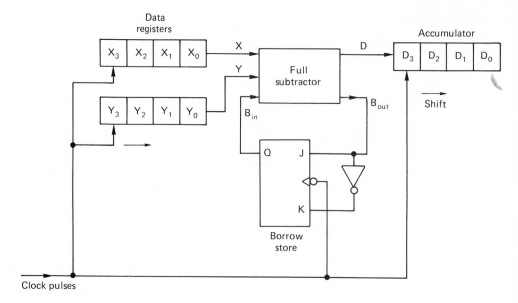

Fig 10.16   Serial subtractor unit

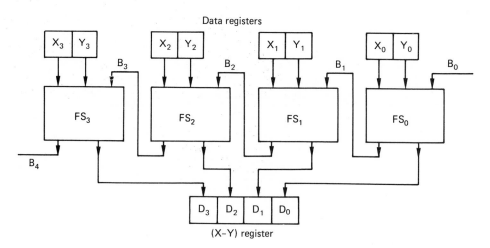

Fig 10.17   Parallel subtractor unit

operations and to use the two's complement representation of signed numbers. A suitable circuit for addition and subtraction is shown in Fig. 10.18.

The most significant bit of the number is used here as the sign bit.

The exclusive-OR gates act as controlled inverters. When the two numbers are to be added, the control line K is set to a '0'. This causes all the XOR gates to allow Y to pass through without change, and the carry input to the least significant adder FA_1 is $C_0 = 0$.

When the two numbers are to be subtracted $(X - Y)$, K is set to a '1'. This changes all Y's to one's complements and makes $C_0 = 1$, thus adding 1 to the one's complements of Y.

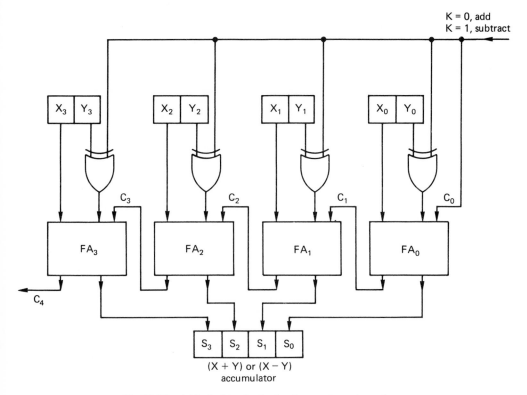

Fig 10.18 Adder/subtractor for two's complement numbers

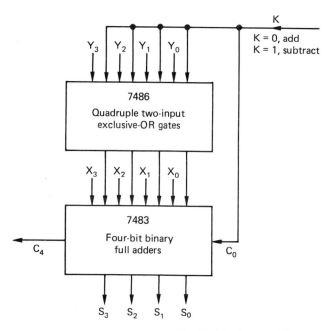

Fig 10.19 Adder/subtractor of Fig. 10.18 using two IC's

Therefore a two's complement of Y is formed before the addition is accomplished. The result appears in the sum accumulator as a two's complement number, with S_3 being the sign digit.

The adder/subtractor of Fig. 10.18 can be implemented with only two integrated circuits, the 7483 four-bit binary full adder and the 7486 quadruple two-input exclusive-OR gates, as shown in Fig. 10.19.

An adder/subtractor for eight-digit signed binary numbers is shown in Fig. 10.20. The most significant digit, bit 7, is now as a sign bit. The circuit can easily be extended to add or subtract *N*-bit binary numbers.

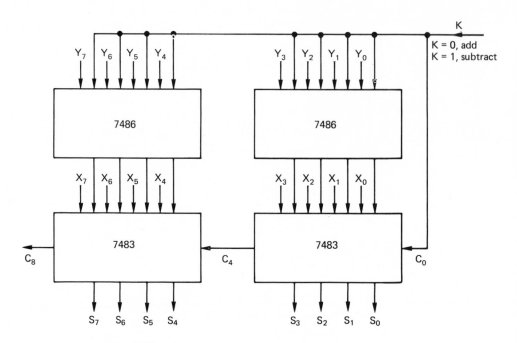

Fig 10.20   Adder/subtractor for eight-digit signed binary numbers

10.6 Serial binary multipliers

Multiplication of binary numbers is discussed in Chapter 2. In this chapter a few possible hardware arrangements are considered. Most of these will be in a block diagram form.

Multiplication by repeated addition

Two numbers, for example 5 and 3, can be effectively multiplied by either adding the number 5 three times or adding the number 3 five times.

A simple circuit to illustrate this process of multiplication by repeated addition is shown in Fig. 10.21.

The number X is loaded to the register A and the down counter is preset to number Y. The counter controls a clock generator which allows the number X to be recirculated Y times while the sum is being gathered together in the product accumulator.

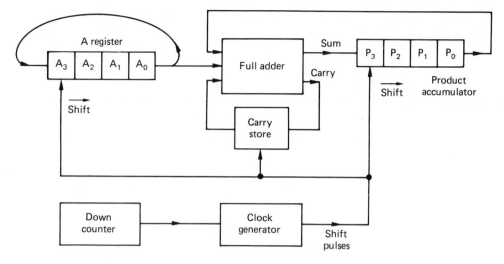

Fig 10.21   Repeated addition multiplier

Multiplication by partial product addition and shifting

Multiplication of two binary numbers can be performed by producing a partial product which, except for the first partial product, is shifted to the left each time it is multiplied by one digit of the multiplier.

An example given here will illustrate design of a suitable circuit. Assuming the two numbers to be multiplied are $X = 1101_2$ and $Y = 1011_2$ (or $13_{10} * 11_{10} = 143_{10}$), then the two binary numbers are multiplied together give the following results:

$$
\begin{array}{rl}
X \quad 1101 & \\
* \quad Y \quad 1011 & \\
\hline
1101 & \text{1st} \\
1101 & \text{2nd} \\
0000 & \text{3rd} \\
1101 & \text{4th} \\
\hline
\end{array}
$$
Partial products

$P \;=\; 10001111_2$

In serial operation, the successive partial products are added to give the required final product. A block diagram of a suitable circuit to perform this operation is shown in Fig. 10.22.

The operation of the circuit is as follows. If the multiplier bit is 1, the AND gate allows the passage for the multiplicand to the full adder, where it is serially added to the contents of the product accumulator, keeping a running sum of the partial product.

If the multiplier bit is 0, the AND gate prevents the transition so that 0 is added and the multiplicand data is recirculated. At the end of each partial product addition the control unit shifts the multiplier one place to the right and the multiplicand one place to the left.

Parallel binary multiplers

The four partial products in the above example can be obtained using four AND gates and shifting Y four times. The partial product can then be added to obtain the final result. For

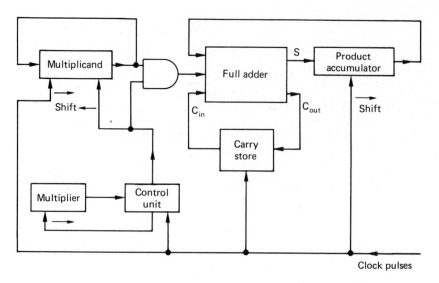

Fig 10.22 Multiplication by partial product addition and shifting

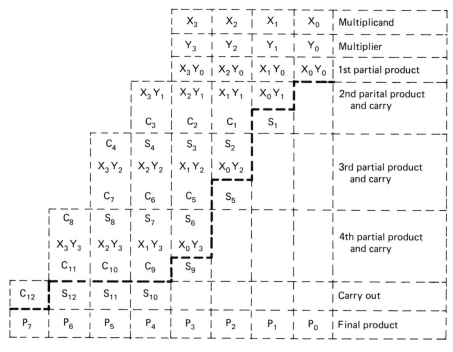

Fig 10.23 Binary multiplication in symbolic form

clarity, the numerical example is repeated using symbolic expressions as shown in Fig. 10.23.

A suitable circuit is then designed using AND gates, half adders and full adders as shown in Fig. 10.24. The multiplier circuit, however, is very uneconomical. It employs 16 AND gates, 4 half adders and 8 full adders. Such a circuit is only acceptable in integrated circuit form.

This circuit can be simplified to a certain extent if each partial product is added as it is formed, as shown in Figure 10.25.

The control unit output is monitored by a down counter which is set to provide the

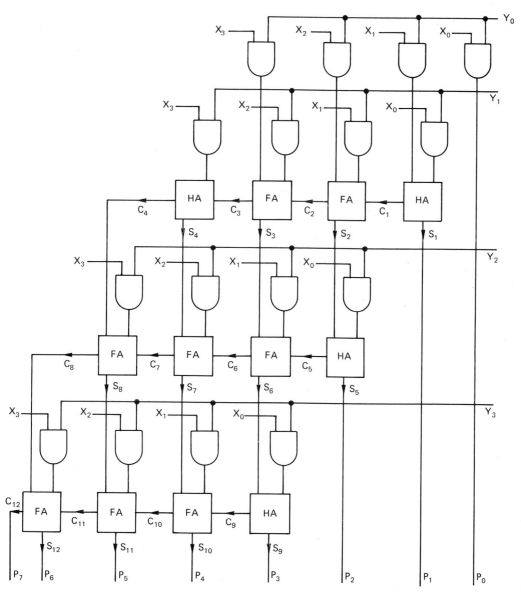

Fig 10.24 Parallel multiplier

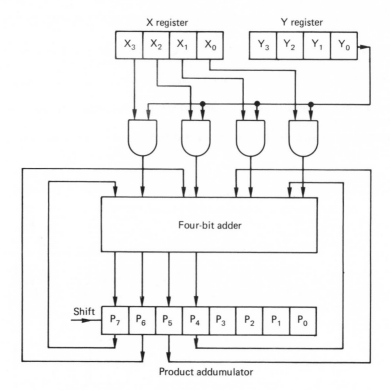

Fig 10.25 Parallel multiplier with partial product shifting

number of pulses equal to the number of multiplier digits. At the end of the multiplication cycle, the final product is available in the product accumulator.

10.7 Binary dividers

In many minicomputers and microprocessor based systems both multiplication and division are performed by software, i.e. programs. Nevertheless, there are logic circuits which perform these operations too. Some hardware solutions for multiplication have already been discussed. Binary dividers can take a number of different forms, but only one or two examples will be mentioned here.

Binary division can simply be performed by repeated subtraction and shifting. The divisor is repeatedly subtracted from dividend until the remainder is less than the divisor. A count is kept of the number of subtractions, which is the required quotient. A functional diagram of a possible system is shown in Fig. 10.26.

Binary multiplication and division by ROM look-up tables

Both multiplication and division can be processed much faster using look-up tables. With the enormous volume storage capabilities of now available semi-conductor memories, the possibility of storing multiplication and division tables is a reality. Tables stored in ROM (Read Only Memory) reduce the size and cost of the system to a minimum. One integrated circuit replaces the entire hardware circuit. This, however, is not the subject of this text.

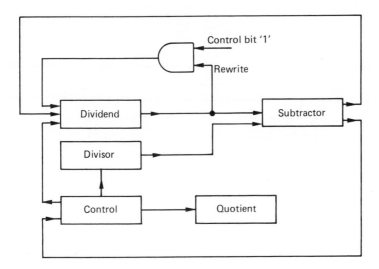

Fig 10.26 Block diagram of a binary divider unit

10.8 Problems

10.1 Explain the difference between a half adder and a full adder.

10.2 Design a full adder circuit using NOR gates only.

10.3 Design a full subtractor circuit using NOR gates only.

10.4 Design a four-bit serial subtractor.

10.5 Explain how a four-bit parallel adder can be modified to act as a combined parallel adder/subtractor. Give step by step explanations for examples of addition and subtraction.

10.6 How can hardware multiplications be carried out using a serial method?

10.7 Describe a method of parallel multiplication and illustrate this with a numerical example.

10.8 Explain briefly how it is possible to implement a hardware divider and give a numerical example.

11

INTEGRATED CIRCUIT
IMPLEMENTATION

11.1 Introduction

This chapter is concerned with the practical considerations involved in implementing a logic design using integrated circuits. This includes consideration of the choice of logic family, scale of integration and package selection. It also contains suggestions on avoiding problems with printed circuit boards and outlines breadboarding methods, test equipment and fault finding techniques.

Chapter 6 looks at the electrical characteristics of the more popular logic families. Chapters 4 and 7 deal with devices, and Chapters 8, 9 and 10 give an idea of how these devices are put together to form useful circuits. Reference has been made in these chapters to a number of available standard integrated circuits containing a few gates, one or two flip-flops or bistables or complete counters or registers. These devices can be combined in an infinite number of ways to provide a required circuit function. The ideal solution is in most cases however to have a single integrated circuit to implement all the specified circuit logical functions. There is no technical barrier to this, only a financial one in a great number of cases. Take, for example, the case of microprocessor based systems. In the majority of such systems, integrated circuits containing relatively few components represent perhaps 75% of the total number of components comprising the system. If only a few systems are to be produced, then clearly there is no point in investing in the production of a single IC to replace the 75% of 'random' logic. Only when tens of thousands of the systems are to be produced per annum will such a solution be cost effective. Small personal computers, which are becoming more widespread nowadays are a good example of this. Whereas previously sales figures could not justify investment in 'custom-built' ICs, the prospect of sales in the tens of thousands and even hundreds of thousands now has made the custom IC a commonplace device in personal and home computers.

Chapter 5 deals with algebraic minimisation techniques to illustrate the importance of reducing the chip count. This is not meant to imply that every circuit design will be based on the use of individual gates. If a specific logic function is required it is usually most convenient to have the function represented in minimal algebraic form. It is possible, however, to implement a logical function directly from a truth table or Boolean expression using data selectors or multiplexers. Other methods of implementation are to use programmable logic arrays or alternatively uncommited logic arrays. These methods will be considered later in this chapter.

11.2 Choice of logic family

To help in the selection of a suitable logic family, Table 11.1 gives typical family characteristics. Each of the logic families mentioned in Table 11.1 are summarised below

Table 11.1 Logic family characteristics

Family	Propagation delay	Power dissipation	Noise immunity	Fan-out	Supply voltage	1 level	0 level
LSTTL	8 ns	2 mW	0.3 V	10	5 V	3.3 V	0.2 V
TTL	10 ns	10 mW	0.4 V	10	5 V	3.3 V	0.2 V
PMOS	100 ns	0.2 mW	1.0 V	50	−20 V	−11 V	−3 V
NMOS	50 ns	0.2 mW	1.0 V	50	15 V	3.5 V	0.4 V
CMOS	40 ns	10 nW	4.0 V	<50	3–15 V	4.99 V*	0.01 V
HCMOS	10 ns	10 μW	1.5 V	10	2–6 V	4.95 V*	0.8 V
STTL	4 ns	20 mW	0.3 V	10	5 V	3.3 V	0.2 V
ECL (10K)	2 ns	25 mW	0.125 V	25	−5.2 V	−1.7 V	−0.8 V

* At $V_{cc} = 5$ V

(1) Standard (high speed) TTL – A high speed logic family with fairly high power consumption.
(2) Schottky TTL – A very high speed logic family with high power consumption.
(3) Low power Schottky TTL – A modified Schottky TTL that consumes less power but with increased propagation delay.
(4) PMOS – Fairly slow with high packing density.
(5) NMOS – Reasonable speed with high packing density.
(6) CMOS – A low power logic family with high noise immunity.
(7) HCMOS – A low power, high speed logic family with high packing density.
(8) ECL – An extremely high speed family with high power consumption.

It must be stressed that the values quoted in Table 11.1 are typical only. Each device has comprehensive characteristics detailed in its own data sheet available from the manufacturer. In selecting the optimum method of producing a desired logic function, remember that the choice of logic family is only one of the many considerations. For example, there is no point in immediately selecting the logic family with the highest operating speed and lowest power consumption if the implementation is prohibitively expensive or if supplies can not be guaranteed.

11.3 Scale of implementation

A rough guide to the number of components contained in an integrated circuit is given by the terms Small Scale Integration (SSI), Medium Scale Integration (MSI), Large Scale Integration (LSI) and Very Large Scale Integration (VLSI). The normal gate densities assigned to these are as follows:

SSI	Up to 10 gates per IC.
MSI	10 to 100 gates per IC.
LSI	100 to 10000 gates per IC.
VLSI	Over 10000 gates per IC.

A typical SSI integrated circuit is the 75HC02 quad 2-input NOR gate which contains four NOR gates in one integrated circuit package.

An example of MSI is the 74HC4511 BCD to 7-segment latch/decoder/driver.

An LSI example is an uncommited logic array (see section 11.6), which may contain several thousand two-input gates.

A 16-bit microprocessor is a good example of VLSI.

Implementation of logical functions using small scale integrated circuits are dealt with fully in previous chapters. There are alternative methods of generating logical functions and these will now be considered.

11.4 Multiplexer logic

It is possible to implement a logical function directly from a truth table using a **multiplexer** or **data selector**. A typical multiplexer chip is the 74HC251 8-channel tristate multiplexer illustrated in Fig. 11.1. There are three inputs to the device, giving $2^3 = 8$ possible conditions on the output line Y. Output W is simply the complement of Y. The 'strobe' line is active low and will enable the outputs if set to logic 0. If the strobe is raised to logic 1, it will set both outputs to a high impedance (tristate) condition. The function to be realised is determined by the data inputs D_0 to D_7, and a particular data output is selected by inputs A, B, and C where A corresponds to the least significant bit (LSB). This is shown in the truth table in Table 11.2.

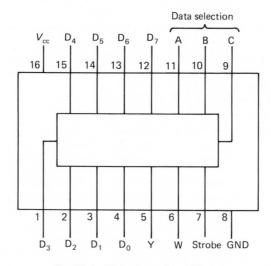

Fig 11.1   Eight channel multiplexer

Table 11.2   Truth table illustrating 74HC251 multiplexer operation (X = don't care)

Inputs				Outputs	
Strobe	Select			Outputs	
S	C	B	A	Y	W
0	0	0	0	D_0	$\overline{D_0}$
0	0	0	1	D_1	$\overline{D_1}$
0	0	1	0	D_2	$\overline{D_2}$
0	0	1	1	D_3	$\overline{D_3}$
0	1	0	0	D_4	$\overline{D_4}$
0	1	0	1	D_5	$\overline{D_5}$
0	1	1	0	D_6	$\overline{D_6}$
0	1	1	1	D_7	$\overline{D_7}$
1	X	X	X	Tristate	Tristate

Table 11.3 Multiplexer desired function

Inputs			Output
C	B	A	Y
0	0	0	0
0	0	1	1
0	1	0	1
0	1	1	1
1	0	0	0
1	0	1	1
1	1	0	1
1	1	1	0

Assume that the function defined in Table 11.3 is required. All that is necessary is to set the D_n input in accordance with the truth table so that

$$D_0 = D_4 = D_7 = 0 \text{ and } D_1 = D_2 = D_3 = D_5 = D_6 = 1$$

Thus the logical function

$$Y = A\overline{B}\overline{C} + \overline{A}B\overline{C} + AB\overline{C} + A\overline{B}C + \overline{A}BC$$

will be realised. This can be achieved as indicated in Fig. 11.2.

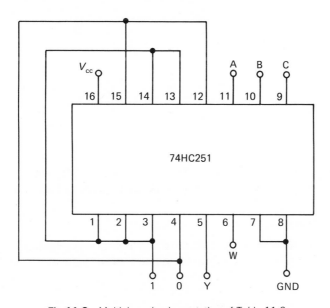

Fig 11.2 Multiplexer implementation of Table 11.3

To implement a four-variable function, the procedure must be modified if the same IC is to be used. Consider the function defined in Table 11.4. This truth table corresponds to the function

$$Y = A\overline{B}\overline{C}\overline{D} + \overline{A}B\overline{C}\overline{D} + AB\overline{C}\overline{D} + \overline{A}\overline{B}C\overline{D} + \overline{A}B\overline{C}D + A\overline{B}\overline{C}D + AB\overline{C}D + \\ + \overline{A}\overline{B}CD + A\overline{B}CD + ABCD$$

This must be rewritten as given in Table 11.5

Table 11.4   Four-input function for multiplexer

Inputs				Output
D	C	B	A	Y
0	0	0	0	0
0	0	0	1	1
0	0	1	0	1
0	0	1	1	1
0	1	0	0	1
0	1	0	1	0
0	1	1	0	0
0	1	1	1	0
1	0	0	0	1
1	0	0	1	1
1	0	1	0	0
1	0	1	1	1
1	1	0	0	1
1	1	0	1	1
1	1	1	0	0
1	1	1	1	1

Table 11.5   Implementation of Table 11.4

Inputs			Output
C	B	A	Y
0	0	0	D
0	0	1	1
0	1	0	$\overline{D}$
0	1	1	1
1	0	0	1
1	0	1	D
1	1	0	0
1	1	1	D

The input connections are now $D_0 = D_5 = D_7 = D$, $D_1 = D_3 = D_4 = 1$, $D_2 = \overline{D}$ and $D_6 = 0$. This can be implemented using the 74HC251 as given in Fig. 11.3.

11.5 Logic arrays

There are a number of devices known collectively as **logic arrays** or **gate arrays**. An example of a **programmable** logic array is the 28-pin bipolar 82S100 PLA, which can be regarded as a number of AND-OR-INVERT(AOI) gates. The 82S100 is designed to implement logical functions expressed in sum-of-product form and allows up to 16 variables to be catered for with up to 48 product terms. For example:

1st variable   15th variable

Second p term

(A.B.C.D. . . M.N.P) + (A.B.C.D. . . M.N.P) + . . . + (A.B.C.D. . . M.N.P.)

1st product (p) term   48th p term

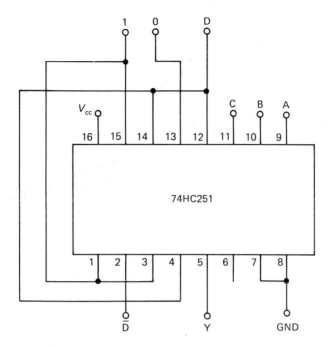

Fig 11.3   Multiplexer implementation of Table 11.5

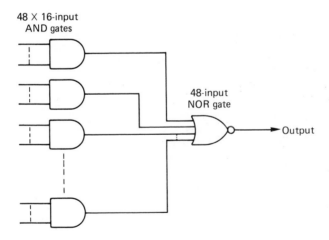

Fig 11.4   1/8th of PLA

The 82S100 is an octal device, which means that eight such expressions can be realised on one PLA chip. Figure 11.4 illustrates how one of these functions is implemented in the PLA.

It is useful to compare the circuit given in Fig. 11.4 with its small scale equivalent. Each 16-input AND gate could be replaced by two eight-input AND gates and one two-input AND gate. There would be, in total, 48 sets of three AND gates, i.e. 144 AND gates.

The 48-input NOR gate could be replaced by 23 three-input OR gates and 1 NOR gate. The total gate count would therefore be $144 + 23 + 1 = 168$ gates.

In terms of ICs:

96	8-input AND gates	96
48	2-input AND gates	12
23	3-input OR gates	8
1	NOR gate	1
		117

This, of course, assumes that all terms are used. In practice this is untypical, and a more realistic estimate is that one PLA of this type replaces 20 SSI integrated circuits. On this basis, the PLA version would probably be double the cost of its SSI equivalent. At first glance, therefore, it seems to be more expensive to use the PLA than random, (discrete) logic. It is necessary to look at the entire production process to see why there is an economic advantage in using a PLA of this type. The manufacturing process costs can be summarised as follows.

(1) Components cost.
(2) IC inspection and testing costs.
(3) Circuit board manufacture, testing and reworking costs.
(4) Labour costs.

The component cost has already been shown to be higher for the PLA than its SSI equivalent. Inspection and testing of ICs is cheaper for the PLA than its equivalent in, say, TTL. The PLA IC has more pins than any of the TTL ICs, but when the number of ICs is taken into account the saving is considerable. Printed circuit board (PCB) costs are a major area of saving if PLAs are used, because:

(a) there is a smaller PCB area because of fewer components;
(b) PCB testing is cheaper for PLAs because of the smaller PCB resulting from point (a).
(c) reworking on PCBs normally required because of incorrectly inserted ICs and bad soldering is reduced. The smaller PCB with less plated holes, soldered connections and ICs that results from using PLAs gives a major saving in this area.

The manufacturer of the PLA described claims that a 55% saving over an equivalent low power Schottky implementation is possible.

A point to consider is that the PLA has to be programmed to generate the appropriate functions. This is done by using a suitable programming device which blows polysilicon fuses at selected points on the IC. It is also possible to implement sequential logic functions using a device known as a **programmable logic sequencer** (PLS). The PLS permits outputs which are a function of the inputs and the current outputs. Counters and binary sequencers can be constructed using a PLS.

11.6 Uncommitted logic arrays

When mass produced circuits for specialist applications are required in high volumes, a 'custom' LSI circuit can be designed and manufactured to suit. Custom circuits will realise a solution with a minimum number of ICs and minimum power consumption for a particular application, and will also maintain design security. Unfortunately, for smaller volumes the cost of custom designs is too high. An **uncommitted logic array** (ULA) is a compromise solution which will provide many of the features of custom circuits at a much lower cost. The basis of ULA design is a standard chip containing an array of components which are initially 'uncommitted', which simply means that the final aluminising layer is absent. The user specifies to the ULA manufacturer the required pattern of component interconnections, which is used to produce a corresponding mask. The mask is used to apply the final metallising layer to the ULA chip, producing the required circuit function. In some instances the end user can install equipment to permit the full design and layout to be carried out on his own premises, requiring only the metallisation to be carried out by the ULA manufacturer.

ULAS can be obtained in a variety of technologies including bipolar, MOS and CMOS. ULAs become cost effective when more than 5000 circuits per annum are required. Lower volumes will, in most cases, be cheaper using conventional SSI and MSI technologies. Volumes higher than 100 000 per annum will probably be cheaper using a full custom designed circuit.

11.7 Integrated circuit packaging

The most common package is the dual-in-line (DIL) which has been illustrated in several places throughout the book. These are available in a large number of sizes, 14- and 16-pin being the most common for small and medium scale integration. Other sizes include 18-, 20-, 24- and 40-pin. Most microprocessor chips are in 40-pin packages. A standard pin-numbering system is used on all DIL packages, as illustrated in Fig. 11.5 for 14-pin, 16-pin and 40-pin packages.

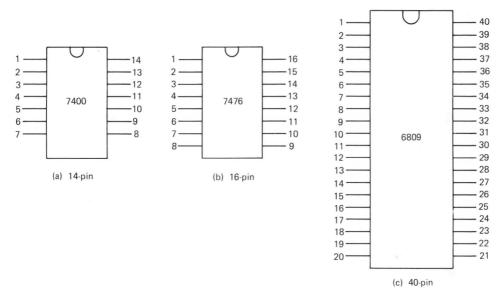

(a) 14-pin   (b) 16-pin   (c) 40-pin

Fig 11.5 Dual-in line pin numbering

Pin 1 is always the end pin below the notch in the package if the notch is positioned at the extreme left hand side when viewed. The IC description should now be printed the correct way up on the package, as illustrated in Fig. 11.6 for a LS TTL 7404 hex inverter.

The pin numbers increase around the right hand side of the package. The package is usually plastic, and will normally operate satisfactorily in the range 0°C to + 70°C which is the normal commercial temperature range. Ceramic packages are also available.

Another common package is the flat pack, illustrated in Fig. 11.7. This package is useful if the IC has to fit into a low profile case.

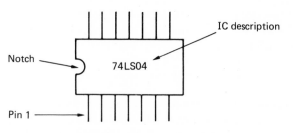

Fig 11.6   DIL pin identification

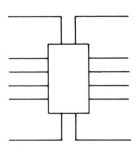

Fig 11.8   To 5 can

Fig 11.7   Flat pack

The TO5 can is illustrated in Fig. 11.8.

Both the TO5 and flat pack will operate over the military temperature range of $-55°C$ to $+125°C$.

A difficulty with Dual-in-Line Integrated Circuits is that precision insertion into holes in the Printed Circuit Board (PCB) are required. Surface Mounted Devices (SMD) mount directly onto pads on the PCB. Most of the devices described in this book are available in Surface Mount Technology versions.

One manufacturer claims that a 70% reduction in board area is possible using SMD devices rather than conventional Dual-in-Line packages. In addition, the use of SMD devices makes design easier to implement, increases reliability and reduces manufacturing costs.

11.8 Breadboarding

Before embarking on final circuit implementation involving a printed circuit board, the circuit should first be tested on a breadboard. A **breadboard** is a unit which allows integrated circuits to be connected together without any soldering, and may include switch inputs, light emitting diodes, seven-segment displays and suitable power supplies. The breadboard should also permit discrete components to be inserted. A fairly basic breadboard unit layout is given in Fig. 11.9.

In this unit, a 5 V power supply is built in and power rails are laid out on the board so that power can be conveniently obtained for an IC inserted in any part of the breadboard. Interconnecting wires can be inserted into spring sockets which connect to IC pins.

The design illustrated does not include switch inputs or LEDs. A convenient method of providing the necessary inputs is to use dual-in line switches which can be inserted into the breadboard in the same way as an integrated circuit. A suitable double-throw IC switch is shown in Fig. 11.10.

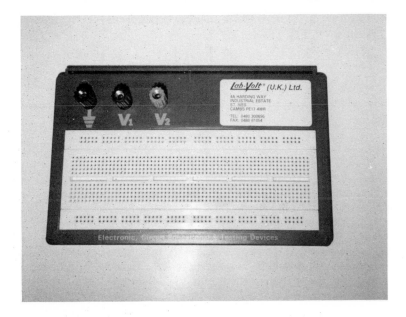

Fig 11.9 Solderless breadbording unit (Courtesy of Lab Volt (UK) Ltd)

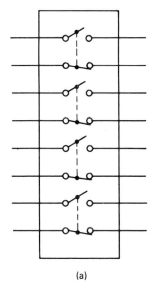

(a)

Fig 11.10 Single-pole double throw (SPDT) dual-in-line switch

Input logic 1 and 0 can be derived via these switches from the V_{CC} and GND supplies respectively, as shown in Fig. 11.11.

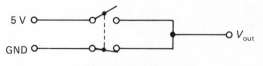

Fig 11.11   Switched inputs

The switches take up valuable space on the breadboarding unit, so if rigidly fixed switches are available these are preferable. The DIL switches will need to be debounced, and this can be carried out by the method given in Section 7.3. Again this will take up space, so it is better to obtain a unit with built-in debounced switches.

Output indication will also be required. This can be provided by means of light emitting diodes (LEDs). Care must be taken not to exceed the maximum current permitted in LEDs, however, and so resistors should be inserted in series with them. Discrete resistors can be used, or alternatively a convenient method is to use resistors in IC form, as shown in Fig. 11.12.

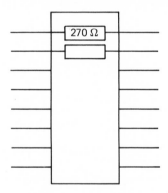

Fig 11.12   DIL resistor IC

Once more this will take up breadboard space, although some breadboards have LEDs and current limiting resistors as standard. A breadboarding unit which has input switches, output LEDs and a built in power supply is given in Fig. 11.13.

An additional requirement will be a clock circuit to produce square waves at a number of different frequencies, for example 1 Hz, to allow a step-by-step examination of circuit operation, and a higher frequency of, say, 100 kHz. If these are not provided they can be derived from a simple circuit using a 555 timer which is easily available and cheap. A suitable circuit is given in Fig. 11.14. For $R_1 \gg R_2$ T = 1.386 $R_1 C_1$, therefore if R_2 = 1 kΩ, R_1 = 68 kΩ and C = 10 μF f = 1 Hz (C = 100 pF for 100 kHz).

For more complex circuits, perhaps involving microprocessors and their peripheral chips, a more professional breadboarding technique is to use wire-wrapping. Wire-wrapping is a method of interconnecting components without soldering by wrapping a wire around a special pin. These pins are on purpose designed wire-wrap IC sockets or other component mounting devices. A typical wire-wrap board is shown in Fig. 11.15.

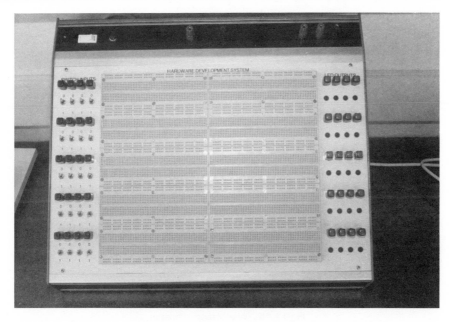

Fig 11.13   Digital circuit breadbording unit

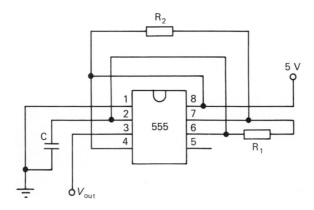

Fig 11.14   Breadboard clock circuit using 555 timer

11.9 Printed circuit board implementation

Since its introduction in the 1940's, the printed circuit board (PCB) has become the standard method of mounting and interconnecting electronic components. A design may be implemented on a single PCB, as in Fig. 11.16, or may consist of a number of boards on a rack system which plugs into a common 'mother board'. The types of board are:

(1) Single-sided.
(2) Double-sided.
(3) Multi-layer.

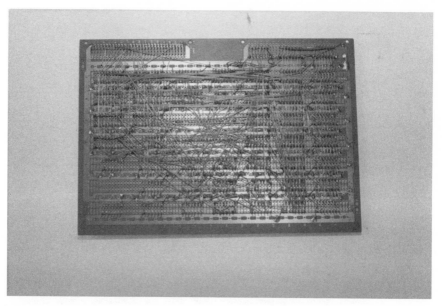

Fig 11.15 Wire wrapped boards

Fig 11.16 Printed circuit board

Single-sided PCBs have conducting tracks on one side of the board only. Double-sided PCBs have conducting tracks on both sides, with connections made from one side of the board to the other by either pins or through-hole copper plating. The major problem here is ensuring registration between tracks on opposite sides of the PCB, i.e. making sure that a conducting track on one side of the PCB is positioned correctly with respect to a track on the other side, so that a through board connection can be directly made. Multi-layer PCBs permit cross-over connections on one side of the board, with the penalty of a more complex production process.

Computer Aided Design (CAD) of PCBs is now a standard method of designing component interconnection paths and track layout but involves an intial capital outlay on hardware and software. Some fairly inexpensive packages for PCB CAD are available for the popular micro and minicomputers.

11.10 Test equipment

If faults are encountered during the breadboarding phase, or if faults arise with the circuit implemented on a printed circuit board, it is essential to have the appropriate fault-finding equipment readily to hand. A number of these tools will now be considered.

A **logic probe** is an important tool in detecting faults. A good logic probe will have the ability to distinguish between a logic '0', a logic '1', and an open circuit or 'floating' input,

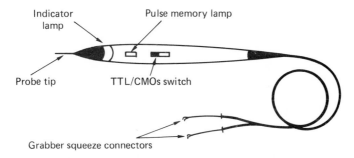

Fig 11.17   Logic probe (Courtesy of Hewlett-Packard)

on all of the popular logic families including TTL and CMOS. This is done by means of an indicator lamp which can either be *OFF* (out), or *ON* (bright), to represent a logic 0 and a logic 1 respectively, or *dim* to represent a floating level. Pulses or clock waveforms are represented by a flashing indicator lamp. Short duration pulses are 'stretched' to a length that can be observed on the lamp by a circuit built into the probe. A typical logic probe is shown in Fig. 11.17. This particular probe has 'grabber squeeze' connecters to permit connection to the power supply of the circuit under test or to a regulated d.c. power supply. The supply voltage can be in the range 4.5 V–15 V for TTL operation and 3 V–18 V for CMOS operation.

A switch on the probe permits selection of TTL or CMOS.

The logic threshold ranges are given in the following table, which assumes 5 V power supplies.

	Logic 1	Logic 0
TTL	1.8–2.4 V	0.5–1.0 V
CMOS	3.0–4.0 V	1.0 V–2.0 V

Consider TTL as an example. Above the logic 1 threshold the lamp is bright, below the logic 0 threshold the lamp is out, and between the two thresholds the lamp is dim. When the probe is not connected to the circuit under test, the lamp is dim. Contact with a 1 or 0 will cause the lamp to go bright or out as appropriate. This makes the task of fault finding on breadboards or printed circuit boards more straightforward. In addition, the probe will detect pulses down to 10 ns input pulse width. The maximum pulse repetition frequency for the probe illustrated is 80 HMz for TTL and 40 MHz for CMOS. It also has a pulse memory lamp which latches the first entry into a valid logic level. An example of the use of a logic probe is given in Fig. 11.18.

The output from the final (OR) gate should be a logic 0, but in fact appears as a logic 1 due to the open circuit on IC_3 which occurs between the gate output on the chip and the IC_3 output pin. The logic probe shows that a valid logic 0 appears at nodes A and B, but

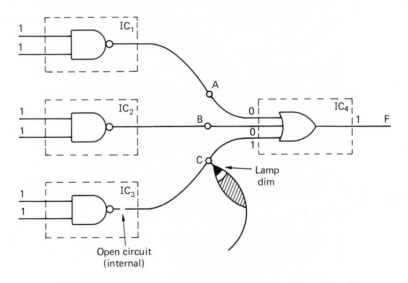

Fig 11.18   Using a logic probe to detect an internal open circuit

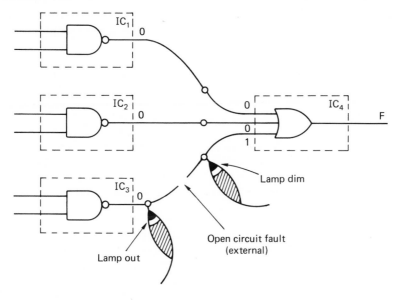

Fig 11.19   Detecting an external open circuit using a logic probe

indicates a 'bad' logic 1 level (floating) at node C caused by the open circuit. Applying the probe to the output of the faulty NAND gate will give the same indication. If the fault is external, as shown in Fig. 11.19, then the output pin of the NAND gate in IC_3 will give a valid 0 on the probe, whereas it will indicate a 'bad' 1 or floating level on the input pin to IC_4.

To correct the fault in Fig. 11.18, IC_3 must be replaced. To correct the fault in Fig. 11.19, the connection between the output of IC_3 and the input of IC_4 must be repaired.

A simple example of the use of the logic probe for pulse detection is in the asynchronous self-resetting counter shown in Fig. 11.20.

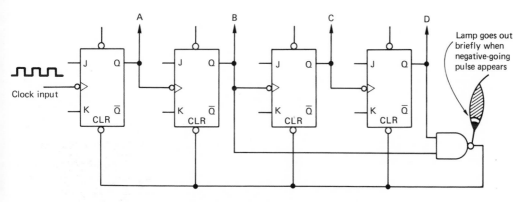

Fig 11.20   Pulse detection using a logic probe

When the count of 10 appears (albeit briefly) on the counter, the NAND gate receives two logic 1 inputs which produce a resetting 0 pulse output to the active low CLEAR inputs on the JK bistables. The outputs A, B, C, D all go to logic 0 and the NAND gate output returns to logic 1. The period of time for which the clear inputs are equal to 0 is extremely short, but will be detected by the probe.

Another useful item of digital test equipment is the **logic pulser** illustrated in Fig. 11.21.

A logic pulser generates controlled pulses which can be applied to a circuit to stimulate a response. It is able to generate or sink sufficient current to overcome existing logic levels at its point of application. The frequency of the pulses can be altered by operating the frequency control switch on the pulser. The number of pulses fed out can also be controlled. An indicating lamp allows the user to monitor the output mode. The amplitude of the pulses generated is determined by the power supply to the pulser. For use with TTL, the logic 0 voltage is less than 0.8 V and the logic 1 voltage is greater than 3 V, assuming a 5 V supply. For CMOS, the logic 0 voltage is less than 0.8 V and the logic 1 voltage is greater than 3 V, assuming a 5 V supply. Also the logic 0 voltage is less than 0.5 V and the logic 1 voltage will be at least equal to the pulser supply voltage minus 1 volt, up to a maximum of 15 V. If a single pulse is required, then this can be achieved by pushing the frequency control switch once. The number of times the switch is pushed determines the output mode, allowing continuous outputs at 1 Hz, 10 Hz, and 100 Hz; also a 1 second burst of 100 pulses every 2 seconds, and a 0.1 second burst of 10 pulses once each second. The pulser can be used to help detect open and short circuits, including solder bridges, and V_{CC} and ground faults, by injecting pulses into the circuit and monitoring the effect with a logic probe or other item of test equipment. An example of the use of the logic pulser together with a logic probe is shown in Fig. 11.22 and a photograph in Fig. 11.23 shows it in use with a current tracer.

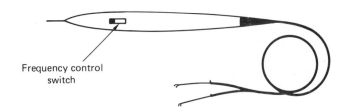

Frequency control
switch

Fig 11.21   Logic pulser

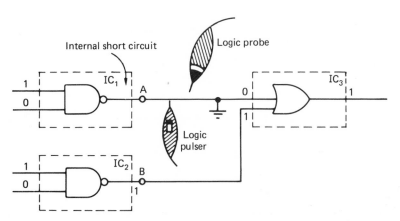

Fig 11.22   Use of logic pulser to detect a short circuit to ground

The output from the final NAND gate located in IC_3 should be a logic 0 but is detected by the logic probe to be a 1. Node B also indicates a logic 1 on the probe, but node A indicates a 0. Is the NAND gate in IC_1 faulty and giving out an incorrect logic level, or is there an earth fault? Applying the pulser and probe as shown will distinguish between these two possibilities. A pulse from the pulser will not be detected on the probe, implying a short circuit to ground. The earth fault detected could be internal to IC_1 or IC_3, or could be external as shown in the diagram. The pulser and probe could similarly be used to indicate the presence of a short circuit to V_{CC}.

Another application for the pulser is to provide a clocking facility for counters, shift registers and bistables. The pulser can be applied to the clock input, and either single pulses or pulse trains can be used to clock the device at a rate which permits circuit operation to be monitored.

A third item of fault finding equipment is the **current tracer**. A typical current tracer is shown in Fig. 11.23.

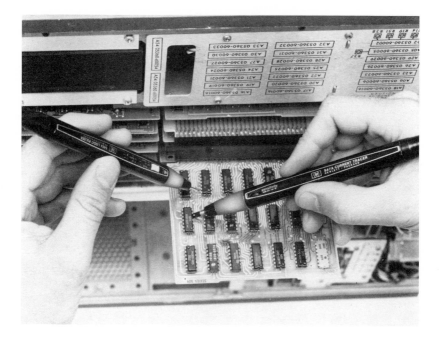

Fig 11.23   Current tracer (Courtesy of Hewlett-Packard)

The current tracer is used to determine the position of low impedance faults by detecting the magnetic field produced by current pulses which originate either in the circuit under test or from a logic pulser. An indicator lamp is lit when the pulses are detected. Probe sensitivity can be adjusted by means of a sensitivity control on the current tracer within the range 1 mA to 1A. The principle of operation of the current tracer is that when a low impedance fault has been detected, it can be used to trace the path of the fault current to locate the fault node. An example of the use of a current tracer in locating a ground fault is shown in Fig. 11.24.

A **logic clip** is designed to clip on to a dual-in-line integrated circuit in 'clothes peg' fashion, and permits the logic states of all pins on the IC under test to be monitored simultaneously. This is achieved by having two rows of LEDs which indicate 1 and 0 levels on the corresponding pins. Some logic clips permit transparent IC pin connection diagrams to facilitate identification. The use of a logic clip in conjunction with a logic pulser enables gate functions to be checked against a truth table, by applying the pulser to gate inputs and monitoring outputs on the logic clip LEDs. A logic clip is shown in Fig. 11.25.

A **logic comparator** is used to compare a suspect IC with a reference IC of the same type. A double row of LEDs indicates whether nodes on the reference IC are at the same logic level as the IC undertest. A difference will cause the LED corresponding to the node in question to light. Power for the comparator is picked up from the power supply to the tested IC. A typical logic comparator is shown in Fig. 11.26.

Logic analysers are more sophisticated digital fault finding instruments used with complex circuits. They have the ability to monitor and record in memory a sequence of logic levels at a number of different points or channels. The analyser illustrated in Fig. 11.27 incorporates an oscilloscope to permit the display of a number of channels of data. Data and address buses in microprocessor systems can be monitored using this device.

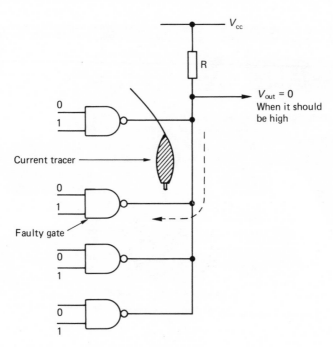

Fig 11.24   Use of a current tracer to locate the path of a fault in a wired AND configuration of open-collector NAND gates

Fig 11.25   Logic clip (Courtesy of Hewlett-Packard)

Fig 11.26   Logic compator (Courtesy of Hewlett-Packard)

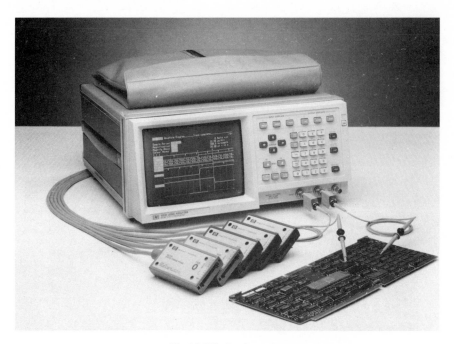

Fig 11.27 Logic analyser

A signature analyser again is designed for use in more complex logic circuits, and is especially useful in tracing faults in microprocessor systems. It is used more frequently by engineers in the field than in the laboratory, and works on the basis of representing a sequence of pulses by a unique 'signature'. The circuit being tested is fed with a test routine which generates the signatures on circuit nodes; these are compared with the expected signature. Any differences will indicate a fault.

11.11 Problems

11.1 What is the most common integrated circuit package in use?

11.2 Which logic family would you choose if gate propagation delay must be less than 3 ns?

11.3 Quote examples of actual chip numbers with descriptions for (a) SSI, (b) MSI, (c) LSI, (d) VLSI.

11.4 Design a circuit to implement the function given in the truth table below using the 74HC251 multiplexer.

Input			Output
A	B	C	Y
0	0	0	1
0	0	1	0
0	1	0	1
0	1	1	1
1	0	0	1
1	0	1	1
1	1	0	0
1	1	1	1

11.5 Explain what is meant by a 'logic array' or 'gate array'. What are the advantages of implementing logical functions using these devices?

11.6 Under what circumstances does it become economical to implement logic circuits using uncommited logic arrays?

11.7 What prototyping methods are in common use in the design and testing of digital circuits?

11.8 Explain the advantages of using printed circuit boards and describe the different types available.

11.9 Explain the operation of the logic probe and show how it could be used to detect: (a) Open circuits; (b) Short circuits to ground; (c) Short circuits to V_{CC}.

11.10 Explain the use of the logic pulser and give an example of its application in fault finding in digital circuits.

11.11 Explain the circumstances in which a current tracer would be used to detect a fault.

11.12 Describe the use of a logic clip.

11.13 Explain the use of the logic comparator in locating a fault in a suspect integrated circuit.

11.14 Under what circumstances would a logic analyser and signature analyser be used?

12

CODES AND CODING CIRCUITS

12.1 Introduction

Chapter 2 introduces the concepts of binary codes. A large number of codes are used in digital systems and computers. Each one has its own specific reason for being used in a particular situation. Common reasons for code selection are:

(1) Simplicity of implementation.
(2) Reduction of errors.
(3) Minimisation of hardware required.
(4) Minimisation of power requirements.

This chapter investigates some of the more commonly used codes and looks at some applications and associated circuits.

12.2 Address representation in computers

Pure binary numbers are frequently used simply to represent numbers and nothing more. For instance, in a microcomputer system store locations are accessed by referring to a unique memory address which is a bit pattern of, say, 16 bit wordlength. In that case addresses between $0000\ 000\ 0000\ 0000_2$ and $1111\ 1111\ 1111\ 1111_2$ can be used, corresponding to the range 0_{10} to $65\ 535_{10}$. Thus with 16 bits a total of $65\ 536_{10}$ memory locations can be accessed.

To actually do this, address decoding must be carried out. (There are a number of methods of designing address decoding circuits for computers. The aim here is merely to introduce the basic concept. The reader should refer to an appropriate text to design a practical decoding circuit.)

Sixteen bits could be decoded to provide 65 536 different address select lines, as shown in Fig. 12.1. Alternatively, the number of select lines could be reduced as in Fig. 12.2. to give a total of 512 store selection lines.

With this system a coincident address selection process allows each of the 65 536 addresses to be accessed.

The choice of decoder would depend upon the nature of the store in use, but clearly the decoding system of Fig. 12.2 is more economical in terms of total select lines than that

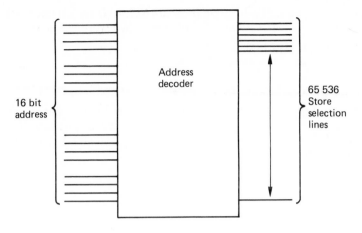

Fig 12.1

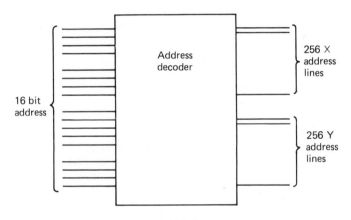

Fig 12.2

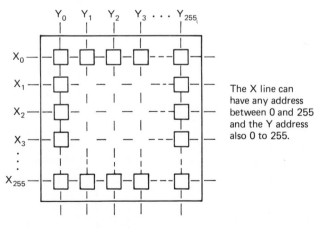

Fig 12.3   The X line can have any address between 0 and 255 and the Y address 0 to 255

depicted in Fig. 12.1. With the first system there are 65 536 (2^{16}) select lines; with the second system there are 512(2×2^8) select lines. In general, with a wordlength of N bits:

address select lines for first system = 2^N

address select lines for second system = $2 \times 2^{\frac{N}{2}}$

12.3 Operation codes in computers

Another, similar, example of the use of a pure binary code to represent information is in a central processing unit (CPU) of a computer or microprocessor. The specific function that is to be carried out is represented by an operation code (op code). For example, typical functions or operations are:

ADD
SUBTRACT
SHIFT RIGHT
SHIFT LEFT
COMPLEMENT
AND
INCLUSIVE-OR
EXCLUSIVE-OR

Each of these has to be represented in code by a bit pattern that uniquely identifies the particular function that has to be carried out in the CPU. For instance:

ADD could be coded as $0000\ 0000_2$
SUBTRACT $0000\ 0001_2$
SHIFT RIGHT $0000\ 0010_2$

and so on to:

EXCLUSIVE-OR $0000\ 0111_2$

Clearly the number of operations is limited by the number of bits available. In the above example with 8 bits, it is possible to have 2^8 or 256 operation codes. In practice it may be necessary to compromise in the number of functions. For example with a 16 bit wordlength the designer has to decide how to split a 16 bit instruction word into an op code and address. He could choose 6 bits for the op code, giving 64 (2^6) functions leaving 10 bits for the address. This would mean that 2^{10} (1 K) of store could be directly accessed. If more

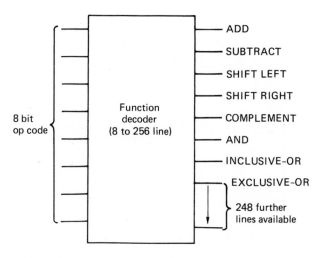

Fig 12.4

directly addressable memory was required, then the number of address bits could be increased – however, this would lead to a reduction in the number of bits representing the op code and, of course, a reduction in the number of operations. Alternatively, extra bytes or words could be added. Decoding op codes can be carried out by a process similar to that used for address decoding as depicted in Fig. 12.1. The eight functions listed previously could be decoded as shown in Fig. 12.4.

Example 12.1
Design a 4-16 line decoder using (a) a diode matrix; (b) simple logic gates.

Solution
See Fig. 12.5.

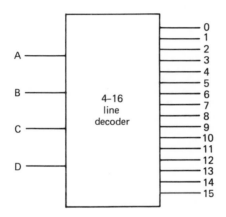

Fig 12.5

The truth table is given in Table 12.1.

Table 12.1

A	B	C	D	0	1	2	3	4	5	6	7	8	9	10	11	12	13	14	15
0	0	0	0	1	0	0	0	0	0	0	0	0	0	0	0	0	0	0	0
0	0	0	1	0	1	0	0	0	0	0	0	0	0	0	0	0	0	0	0
0	0	1	0	0	0	1	0	0	0	0	0	0	0	0	0	0	0	0	0
0	0	1	1	0	0	0	1	0	0	0	0	0	0	0	0	0	0	0	0
0	1	0	0	0	0	0	0	1	0	0	0	0	0	0	0	0	0	0	0
0	1	0	1	0	0	0	0	0	1	0	0	0	0	0	0	0	0	0	0
0	1	1	0	0	0	0	0	0	0	1	0	0	0	0	0	0	0	0	0
0	1	1	1	0	0	0	0	0	0	0	1	0	0	0	0	0	0	0	0
1	0	0	0	0	0	0	0	0	0	0	0	1	0	0	0	0	0	0	0
1	0	0	1	0	0	0	0	0	0	0	0	0	1	0	0	0	0	0	0
1	0	1	0	0	0	0	0	0	0	0	0	0	0	1	0	0	0	0	0
1	0	1	1	0	0	0	0	0	0	0	0	0	0	0	1	0	0	0	0
1	1	0	0	0	0	0	0	0	0	0	0	0	0	0	0	1	0	0	0
1	1	0	1	0	0	0	0	0	0	0	0	0	0	0	0	0	1	0	0
1	1	1	0	0	0	0	0	0	0	0	0	0	0	0	0	0	0	1	0
1	1	1	1	0	0	0	0	0	0	0	0	0	0	0	0	0	0	0	1

(a) The diode matrix solution would be as in Fig. 12.6.

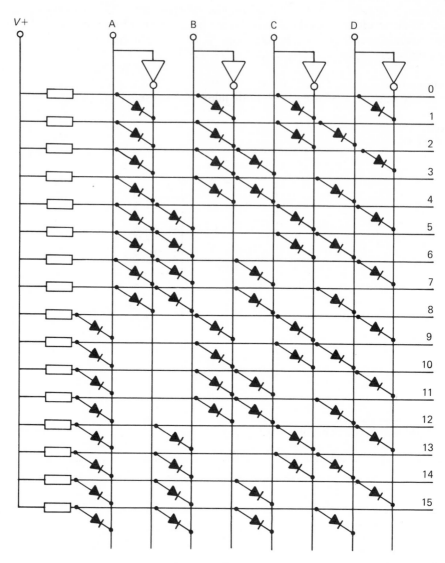

Fig 12.6

The inputs are applied to A, B, C, D. A 0 applied to an input will drag down to 0 the cathodes of any diode connected directly to that input. Cathodes connected via the inverter will be maintained at 1. A diode with its cathode at 0 will drag down to 0 the output to which it is connected. Thus to maintain an output at 1, the cathodes of all the diodes connected to that output must be at 1. For instance if A = 0, B = 1, C = 1, D = 1, then only output line 7 meets the condition of having all diode cathodes at logic 1. To design the decoder simply connect diodes to correspond to this condition.

(b) The logic gate solution for this problem is given in Fig. 12.7.

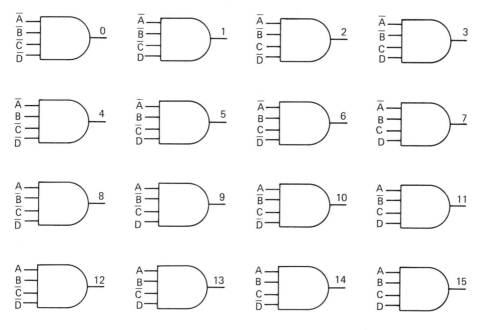

Fig 12.7

Of course this solution could be implemented using alternative types of gates following the design procedure previously outlined. A 4–16 line decoder is available on a single integrated circuit.

12.4 Analogue representation

Another example of where binary numbers are used as a form of code is in digital to analogue (D/A) and analogue to digital (A/D) conversion. In D/A conversion a binary number is converted to a proportional output voltage. In A/D conversion the reverse process takes place. Thus patterns of 0s and 1s represent actual voltage values.

12.5 Encoding circuits

The decoders previously examined convert a binary pattern into another form. The reverse of this process is known as encoding. An example follows which illustrates one type of encoder and shows how it may be implemented using logic gates.

Example 12.2
Design an encoder using logic gates to convert the decimal numbers 1 to 9 to binary.

Solution
See Fig. 12.8 for a block diagram and Fig. 12.9 using OR gates.

12.6 Alphanumeric codes

Example 12.2 gives one method of converting a decimal number into a bit pattern. In order to be able to cater for the full range of characters encountered on a keyboard, including

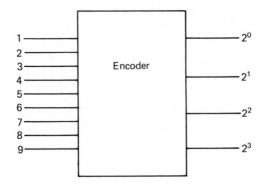

Fig 12.8

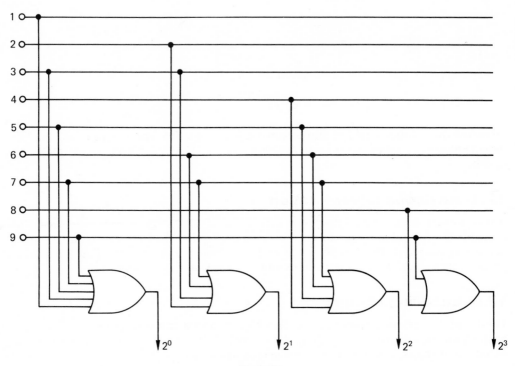

Fig 12.9

decimal numbers, a more detailed code is needed. Codes that represent alphabetic characters and numbers are known as alphanumeric codes. Typical requirements might be:

		number of symbols
Decimal numbers	0–9	10
Upper case letters	A–Z	26
Lower case letters	a–z	26
Other symbols	(*, etc.	15 (say)
Total		77

So in this case, 77 symbols have to be represented by unique bit patterns.

If a 6-bit code were to be used, this would give $2^6 = 64$ patterns, which is insufficient, so 7 bits are needed. This gives $2^7 = 128$ patterns, which in this case would mean that 51 patterns are redundant. However, this gives room for expansion if new characters are to be introduced.

Example 12.3
Design an alphanumeric code to include upper and lower case letters and 15 symbols. Tabulate the code giving hex and octal equivalents.

Solution
Table 12.2(a) shows a possible code for the decimal numbers 0–9 which are represented by the hex values 00 to 09, and the upper case letters A–Z which are represented by hex 0A to 23.

Table 12.2(a)

Character	Binary pattern	Decimal	Hex	Octal
0	00000000	0	00	000
1	00000001	1	01	001
2	00000010	2	02	002
3	00000011	3	03	003
4	00000100	4	04	004
5	00000101	5	05	005
6	00000110	6	06	006
7	00000111	7	07	007
8	00001000	8	08	010
9	00001001	9	09	011
A	00001010	10	0A	012
B	00001011	11	0B	013
C	00001100	12	0C	014
D	00001101	13	0D	015
E	00001110	14	0E	016
F	00001111	15	0F	017
G	00010000	16	10	020
H	00010001	17	11	021
I	00010010	18	12	022
J	00010011	19	13	023
K	00010100	20	14	024
L	00010101	21	15	025
M	00010110	22	16	026
N	00010111	23	17	027
O	00011000	24	18	030
P	00011001	25	19	031
Q	00011010	26	1A	032
R	00011011	27	1B	033
S	00011100	28	1C	034
T	00011101	29	1D	035
U	00011110	30	1E	036
V	00011111	31	1F	037
W	00100000	32	20	040
X	00100001	33	21	041
Y	00100010	34	22	042
Z	00100011	35	23	043

Table 12.2(b)

Character	Binary pattern	Decimal	Hex	Octal
a	0100100	36	24	044
b	0100101	37	25	045
c	0100110	38	26	046
d	0100111	39	27	047
e	0101000	40	28	050
f	0101001	41	29	051
g	0101010	42	2A	052
h	0101011	43	2B	053
i	0101100	44	2C	054
j	0101101	45	2D	055
k	0101110	46	2E	056
l	0101111	47	2F	057
m	0110000	48	30	060
n	0110001	49	31	061
o	0110010	50	32	062
p	0110011	51	33	063
q	0110100	52	34	064
r	0110101	53	35	065
s	0110110	54	36	066
t	0110111	55	37	067
u	0111000	56	38	070
v	0111001	57	39	071
w	0111010	58	3A	072
x	0111011	59	3B	073
y	0111100	60	3C	074
z	0111101	61	3D	075

Table 12.2(c)

Character	Binary pattern	Decimal	Hex	Octal
[	0111110	62	3E	076
]	0111111	63	3F	077
.	1000000	64	40	100
,	1000001	65	41	101
:	1000010	66	42	102
;	1000011	67	43	103
(	1000100	68	44	104
)	1000101	69	45	105
?	1000110	70	46	106
!	1000111	71	47	107
@	1001000	72	48	110
"	1001001	73	49	111
–	1001010	74	4A	112
*	1001011	75	4B	113
&	1001100	76	4C	114

Table 12.2(b) tabulates the code values for the lower case letters a–z which are represented by the hex values 24 to 3D.

Table 12.2(c) lists 15 typical symbols which are represented by the hex values 3E to 4C. A code that has become the standard for computer input/output is the American

Standard Code for Information Interchange, or **ASCII** as it is usually referred to. The ASCII code is basically a 7 bit alphanumeric code with an eighth, parity bit attached. Parity is a method to help detect and possibly correct errors that occur during data transmission. If *even* parity is used, the total number of 1s in any 8 bit pattern in the code is made to be an even number by setting the parity bit to 0 or 1 appropriately. If *odd* parity is used, the parity bit makes the complete pattern odd. The type of parity used will depend upon the system in use; some systems may simply set the parity bit to a permanent 0 or 1 if parity checking is not required. This gives the appearance when checking the hex or actual values that there are a number of different ASCII codes.

One form of ASCII code is shown in Table 12.3.

Table 12.3(a)

Character	Bit pattern	Hex
0	0011 0000	30
1	1011 0001	B1
2	1011 0010	B2
3	0011 0011	33
4	1011 0100	B4
5	0011 0101	35
6	0011 0110	36
7	1011 0111	B7
8	1011 1000	B8
9	0011 1001	39
A	0100 0001	41
B	0100 0010	42
C	1100 0011	C3
D	0100 0100	44
E	1100 0101	C5
F	1100 0110	C6
G	0100 0111	47
H	0100 1000	48
I	1100 1001	C9
J	1100 1010	CA
K	0100 1011	4B
L	1100 1100	CC
M	0100 1101	4D
N	0100 1110	4E
O	1100 1111	CF
P	0101 0000	50
Q	1101 0001	D1
R	1101 0010	D2
S	0101 0011	53
T	1101 0100	D4
U	0101 0101	55
V	0101 0110	56
W	1101 0111	D7
X	1101 1000	D8
Y	0101 1001	59
Z	0101 1010	5A

Table 12.3(b)

Character	Bit pattern	Hex
Space	1010 0000	A0
!	0010 0001	21
''	0010 0010	22
#	1010 0011	A3
£	0010 0100	24
%	1010 0101	A5
&	1010 0110	A6
'	0010 0111	27
(	0010 1000	28
)	1010 1001	A9
*	1010 1010	AA
+	0010 1011	2B
,	1010 1100	AC
.	0010 1110	2E
/	1010 1111	AF
:	0011 1010	3A
;	1011 1011	BB
<	0011 1100	3C
=	1011 1101	BD
>	1011 1110	BE
?	0011 1111	3F
@	1100 0000	C0
[	1101 1011	DB
\	0101 1100	5C
]	1101 1101	DD
↑	1101 1110	DE
←	0101 1111	5F
LF	0000 1010	0A
CR	1000 1101	8D
EOT	1000 0100	84
BELL	1000 0111	87
ESC	0001 1011	1D
DEL	1111 1111	FF

12.7 Binary coded decimal codes

The ASCII code is an alphanumeric code to cater for alphabetic and numeric characters, but frequently there is a need to deal with decimal numbers only. Codes designed for this purpose are known as **binary coded decimal** or **BCD** codes. The concept of a BCD code was introduced in Chapter 2. To define decimal numbers in a binary code requires 4 bits. Of course this gives 16 possible 4 bit patterns that can be selected, to represent only 10 digits (0, 1, 2, 3, 4, 5, 6, 7, 8, 9). The choice of which ten patterns to choose depends upon system requirements. The number of ways of choosing r out of n is given by:

$$\frac{n!}{(n-r)!}$$

In this case $n = 16$ and $r = 10$, giving:

$$\frac{16!}{(16-10)!} = \frac{16!}{6!} = 16 \times 15 \times 14 \times 13 \times 12 \times 11 \times 10 \times 9 \times 8 \times 7$$
$$= 2.9 \times 10^{10}$$

which means that a vast number of BCD codes can be used.

Table 12.4

Decimal	BCD
0	0000
1	0001
2	0010
3	0011
4	0100
5	0101
6	0110
7	0111
8	1000
9	1001

The obvious first choice for BCD representation is to use the pure binary equivalents of the ten decimal digits (Table 12.4). As the binary digits are weighted in the sequence 8421, this particular code is usually referred to as 8421 BCD. This leaves the following 6 redundant bit patterns:

1010	(10)
1011	(11)
1100	(12)
1101	(13)
1110	(14)
1111	(15)

This redundancy can be helpful in error detection. If one bit is corrupted in the BCD equivalent of 9, for instance, then:

9   1001   after corruption   1011

This generates a redundant pattern corresponding to (11_{10}). To detect all the redundant states an error detecting circuit can be used. By inspection, a redundant state occurs when the condition:

$$F = C.D + B.D = D(B + C)$$

occurs where A is the least significant bit. This condition can be simply implemented as shown in Fig. 12.10.

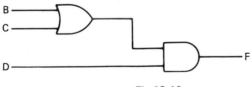

Fig 12.10

In 8421 BCD as described above, the number 479 would be represented as:

0100 ⟍   0111   ⟋1001 = 010001111001
       479

whereas in pure binary, 479 would be represented as:

000111011111

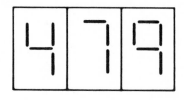

Fig 12.11

It may be required to display the number 479 on seven-segment displays as shown in Fig. 12.11 Each digit has to be addressed separately through decoder/driver circuits, as shown in Fig. 12.12.

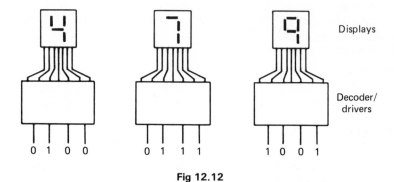

Fig 12.12

As can be seen from Fig. 12.12, each decoder/driver circuit must be fed with the appropriate bit pattern in 8421 BCD format. However, the number 479 may be in pure binary form, that is 000111011111 as above. This means that a decoding circuit must be included to change pure binary into 8421 BCD format to provide the correct form of inputs to the display circuitry.

Example 12.4
Design a circuit to convert 4 bit pure binary into 8421 BCD.

Solution
See Fig. 12.13.

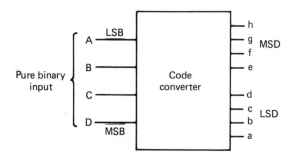

Fig 12.13

Table 12.5

	D	C	B	A	h	g	f	e	d	c	b	a
0	0	0	0	0	0	0	0	0	0	0	0	0
1	0	0	0	1	0	0	0	0	0	0	0	1
2	0	0	1	0	0	0	0	0	0	0	1	0
3	0	0	1	1	0	0	0	0	0	0	1	1
4	0	1	0	0	0	0	0	0	0	1	0	0
5	0	1	0	1	0	0	0	0	0	1	0	1
6	0	1	1	0	0	0	0	0	0	1	1	0
7	0	1	1	1	0	0	0	0	0	1	1	1
8	1	0	0	0	0	0	0	0	1	0	0	0
9	1	0	0	1	0	0	0	0	1	0	0	1
10	1	0	1	0	0	0	0	1	0	0	0	0
11	1	0	1	1	0	0	0	1	0	0	0	1
12	1	1	0	0	0	0	0	1	0	0	1	0
13	1	1	0	1	0	0	0	1	0	0	1	1
14	1	1	1	0	0	0	0	1	0	1	0	0
15	1	1	1	1	0	0	0	1	0	1	0	1

With 4 bits, numbers range from 0 to 15 so 2 digits (for instance 1 and 5) have to be catered for. The truth table, Table 12.5, looks rather formidable if a conventional design routine is to be entered into. By inspection it can be seen that overflowing from 9 to 10 is equivalent to adding 6 to the original number if the output abcdefgh is temporarily regarded as being a pure binary number. That is:

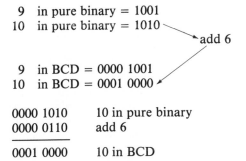

9 in pure binary = 1001
10 in pure binary = 1010 — add 6

9 in BCD = 0000 1001
10 in BCD = 0001 0000

0000 1010	10 in pure binary
0000 0110	add 6
0001 0000	10 in BCD

So all that is necessary is to detect when the binary number is 10 or greater, and add 6.

The binary number will be greater than 10 when D AND B OR D AND C are at logic 1, that is:

$$F = D.B + D.C$$

which looks rather familiar and in fact cropped up when redundancy was being considered.

The problem can be solved as shown in Fig. 12.14. If the pure binary number is less than 10_{10}, then the AND gate outputs are all at logic 0. Thus 0 is added to the binary number as required for this condition. If the pure binary number is 10_{10} or greater, then the AND gates are all enabled and the bit pattern 0110_2 is added to the pure binary number to give the required overflow. The addition can be realised by two parallel adders, one for the *units* and one for the *tens* components of the BCD result. The circuit shown in Fig. 12.14 could easily be extended to deal with pure binary numbers of larger wordlength. Note that the sum S_0, S_1, S_2, S_3 corresponds to abcd, and the sum S_4, S_5, S_6, S_7 corresponds to efgh.

8421 BCD is a straightforward code to use, because weighting is the same as in pure binary and arithmetic operations can accordingly be carried out very simply.

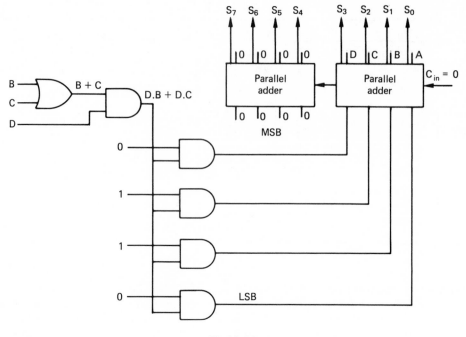

Fig 12.14

Another possible choice of 10 patterns out of the 16 available is shown in Table 12.6. A brief inspection of the code will indicate that each of the binary values have been incremented by three, relative to the pure binary BCD. For this reason it is known as **excess-3** (XS3) BCD. At first glance it seems strange simply to add 3 to each value but there are very sound reasons for doing so. This will be illustrated by means of examples. One point to notice before examining the application of this code is that the code is not weighted as with pure binary BCD, i.e. it is not possible to determine the decimal equivalent of a bit pattern in excess-3 BCD merely by adding up weights. This complicates encoding and decoding circuits but is not an insurmountable problem.

Table 12.6

Decimal	DCBA
0	0011
1	0100
2	0101
3	0110
4	0111
5	1000
6	1001
7	1010
8	1011
9	1100

Example 12.5
Design a logic circuit to convert excess-3 BCD to decimal.
Use NOR logic.

Solution
From the truth table for XS3:

$0 = \bar{C}.\bar{D} \quad = \overline{C + D}$ $5 = \bar{A}.\bar{B}.\bar{C} = \overline{A + B + C}$

$1 = \bar{B}.\bar{C}.\bar{D} = \overline{B + C + D}$ $6 = A.\bar{B}.\bar{C} = \overline{\bar{A} + B + C}$

$2 = A.\bar{B}.C = \overline{\bar{A} + B + \bar{C}}$ $7 = \bar{A}.B.\bar{C} = \overline{A + \bar{B} + C}$

$3 = \bar{A}.B.C = \overline{A + \bar{B} + \bar{C}}$ $8 = A.B.D = \overline{\bar{A} + \bar{B} + \bar{D}}$

$4 = A.B.C = \overline{\bar{A} + \bar{B} + \bar{C}}$ $9 = C.D \quad = \overline{\bar{C} + \bar{D}}$

See Fig. 12.15.

Excess-3 addition

Consider the following addition of two excess-3 numbers:

```
+ 4      0111
+ 3      0110
-----    -----
+ 7      1101
```

The sum appears as 1101 by the normal method of adding binary numbers. The excess-3 equivalent of + 7 is 1010 which is not the result given above. The reason is that adding two excess-3 numbers gives an **excess-6** (XS6) result *not* an excess-3 result. The solution is obtained by subtracting 3 to convert from XS6 to XS3:

```
+ 4          0111
+ 3          0110
-----        -----
+ 7          1101
           - 0011
             -----
             1010   excess-3 seven.
```

This procedure can *only* be followed if the sum of the two excess-3 numbers is less than 10. Consider what happens if this is not the case.

```
+ 6               1001      Note that conveniently a
+ 7               1010      1 is carried out when the
-----             -----     sum is ⩾ 10 because the
+ 13     0001     0011      result is XS6.
         carry
```

In this case a 1 is carried, out giving a total of 0001 0011, i.e. 13 in decimal. The result is therefore correct in 8421 BCD because of the generated carry, as illustrated in Example 12.4. The result, however, is *not* in excess-3, so 3 must be added:

```
0001 0011
     0011
-----------
0001 0110   which is the XS3 equivalent of 13
```

If there is *no* carry then the XS6 result must be converted to XS3 by subtracting 3.
    In summary, to obtain an XS3 BCD result:

    if the sum is < 10, subtract 3.
    if the sum is ⩾ 10, add 3.

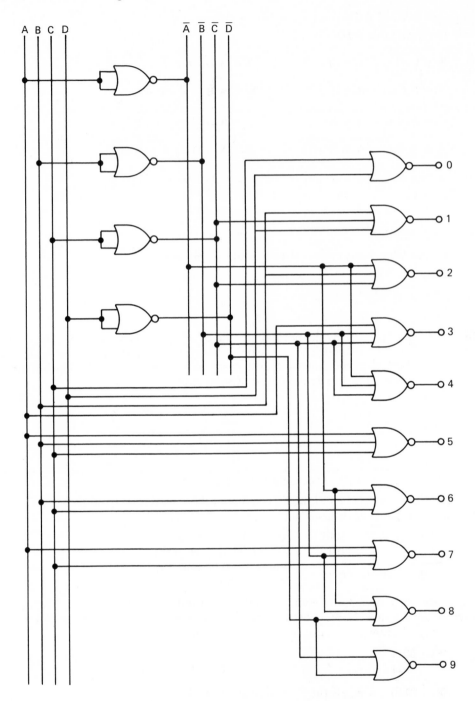

Fig 12.15   Excess-3 BCD to decimal decoder

Example 12.6
Add together the decimal numbers 267 and 149 using XS3 BCD arithmetic.

Solution

Decimal	Excess-3		
267	0101	1001	1010
149	0100	0111	1100
416	1010	0001	0110
	− 0011	+ 0011	+ 0011
	0111	0100	1001

so the result in XS3 BCD is

0111 0100 1001

which corresponds to decimal 416.

Example 12.7
Design a 4-bit XS3 BCD adder.

Solution

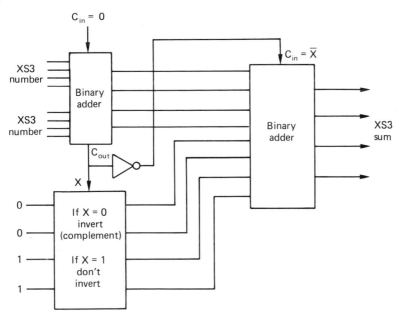

Fig 12.16

An advantage of XS3 BCD is that it is a self complementing code. For example, the XS3 BCD equivalent of 7 is 1010. If the bits are reversed, then 0101 is obtained, which is the XS3 BCD equivalent of 2.

$7 + 2 = 9$

so 2 is the **nine's complement** of 7.

When dealing with decimal numbers it is convenient to use the nine's and ten's complements instead of one's and two's complements when representing positive and negative values. For example, with the number 12,

12_{10} in XS3 BCD ────────────▶ 0100 0101
complement of 12_{10} ───────────▶ 1011 1010

which is 87.

$$12 + 87 = 99$$

This means that the nine's complement of the XS3 BCD number can be obtained merely by finding the one's complement of each 4 bit digit. This is extremely useful when the nine's complement technique is used for subtraction.

Other BCD codes

Aiken codes
The Aiken code is a weighted BCD code (2421) which is self-complementing (see Table 12.7). The Aiken Code follows normal 8421 BCD up to 4, and then jumps to 1011 at 5. Five to nine in Aiken are equivalent to 11 to 15 in 8421 BCD.

Table 12.7

Decimal	Aiken
	2421
0	0000
1	0001
2	0010
3	0011
4	0100
5	1011
6	1100
7	1101
8	1110
9	1111

7421 BCD
The code is tabulated in Table 12.8. The similarity with 8421 BCD means that the code is identical up to 6_{10}. In 8421 BCD, 7_{10} would need to be represented by 0111 which contains three logic 1 levels. To minimise the number of logic 1 levels the 7421 code uses the combination 1001. This means that if a logic 1 corresponds to greater power consumption than a logic 0, the use of 7421 BCD would result in less power consumption than 8421 BCD for that value. If the number of 1s in 7421 BCD are counted then a total of 14 is obtained. With Aiken code the total number of 1s is 20, with obvious disadvantages in terms of power consumption. Although the 7421 code is weighted BCD, the sum of the weights is not 9 and the code is not self complementing.

2421, 5421, 5211 BCD
These three weighted codes are shown in Table 12.9. Although the Aiken code is 2421 as previously described, it is not the only 2421 code possible as the listing in Table 12.9 illustrates.

Table 12.8

Decimal	7421
0	0000
1	0001
2	0010
3	0011
4	0100
5	0101
6	0110
7	1000
8	1001
9	1010

Table 12.9

Decimal	2421	5421	5211
0	0000	0000	0000
1	0001	0001	0001
2	0010	0010	0100
3	0011	0011	0110
4	0100	0100	0111
5	0101	1000	1000
6	0110	1001	1001
7	0111	1010	1011
8	1110	1011	1110
9	1111	1100	1111

Negatively weighted codes

It is also possible to have a code with negative weightings. The example shown in Table 12.10 is of a code with weights $+8$ $+4$ -2 -1. Note that this code is self-complementing.

Table 12.10

Decimal	+8	+4	-2	-1
0	0	0	0	0
1	0	1	1	1
2	0	1	1	0
3	0	1	0	1
4	0	1	0	0
5	1	0	1	1
6	1	0	1	0
7	1	0	0	1
8	1	0	0	0
9	1	1	1	1

12.8 Gray codes

An example of a code which is not BCD is the four-bit Gray code which is shown in Table 12.11. Note how the code is obtained by reflection.

Table 12.11

A	B	C	D		Decimal
0	0	0	0		0
0	0	0	1		1
0	0	1	1		2
0	0	1	0		3
0	1	1	0		4
0	1	1	1		5
0	1	0	1		6
0	1	0	0		7
1	1	0	0		8
1	1	0	1		9
1	1	1	1		10
1	1	1	0		11
1	0	1	0		12
1	0	1	1		13
1	0	0	1		14
1	0	0	0		15

Column D   0 and 1 in the first 2 rows (row 0 and row 1) are reflected to give 1 and 0 in rows 2 and 3. This pattern is repeated in column D.

Column C   The two 0's , two 1's pattern in rows 0 to 3 is reflected to give a two 1's, two 0's pattern in rows 4 to 7 and is repeated throughout column C.

Column B   Four 0's and four 1's in rows 0 to 7 are reflected to give rows 8 to 15.

Column A   This column contains eight 0's and eight 1's, which would be reflected if there were more than 16 rows.

This technique allows a Gray code of any wordlength to be generated.

The Gray code is used particularly in analogue- to-digital conversion where the angular position of a rotating shaft has to be accurately determined. This is achieved by means of a shaft encoder which is attached to the rotating shaft. Normally in the form of a disk or drum, the encoder is coded such that a change in angular position corresponds to a change in bit pattern. The obvious first choice for a code to use is pure binary as this can readily be recognised and decoded. A coded three-bit disk using pure binary is shown in Fig. 12.17. A shaded area refers to a logic 1, an unshaded area to a logic 0.

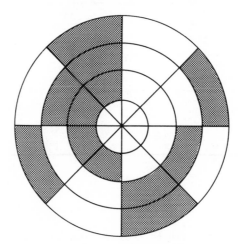

Fig 12.17   Binary coded disk

Consider the change in bit pattern as a four-bit disk rotates from 15_{10} to 0_{10}.

1111
0000

Note that all four bits change. This is fine as long as you can be certain that the read heads will all detect the changes at the same time. But what happens if, perhaps due to head misalignment, they do not? The worst case is considered in Fig. 12.18.

1	1	1	1
0	1	1	1
0	0	1	1
0	0	0	1
0	0	0	0

Fig 12.18 Intermediate state errors

Assuming that the head is misaligned as shown, instead of getting a nice clean change from fifteen to zero the sequence 15, 7, 3, 1, 0 will be detected, giving the intermediate error states of 7, 3, 1. This incorrectly implies that the disk is frantically changing its angular position before reaching zero. This only occurs because of the fact that the code changes by more than one bit at a time. A code with only one bit change at a time would eliminate this. The Gray code is a very commonly used code that possesses this feature.

12.9 Generation of code sequences using counters

Aiken Counters

The Aiken Counter is a synchronous BCD 2421 counter which follows the symmetrical nine's complement code shown in Table 12.12.

Table 12.12 Transition table for BCD 2421 Aiken code

Input pulses	Present state Q_n				Next state Q_{n+1}				Nine's complement
	D	C	B	A	D	C	B	A	
	2	4	2	1	2	4	2	1	
0	0	0	0	0	0	0	0	1	9
1	0	0	0	1	0	0	1	0	8
2	0	0	1	0	0	0	1	1	7
3	0	0	1	1	0	1	0	0	6
4	0	1	0	0	1	0	1	1	5
5	1	0	1	1	1	1	0	0	4
6	1	1	0	0	1	1	0	1	3
7	1	1	0	1	1	1	1	0	2
8	1	1	1	0	1	1	1	1	1
9	1	1	1	1	0	0	0	0	0

Since only ten combinations of the binary digits are used, the following six digits are redundant: 0101, 0110, 0111, 1000, 1001 and 1010. These will be marked with X on the Karnaugh map for the BCD 2421 Aiken code, shown in Figs. 12.19 and 12.20. Reference is made to the J-K output transition table in order to transfer the information required from the output state change maps to the J-K input maps in Fig 12.20.

Having performed the necessary looping, the following expressions are derived for the flip-flop input signals: $J_A = K_A = 1$, $J_B = A + C\bar{D}$, $K_B = A$, $J_C = AB$, $K_C = AB + \bar{D}$, $J_D = C$ and $K_D = ABC$. The counter is then implemented as shown in Fig. 12.21.

Alternative implementation of the counter is given, since this is a classical example where some of the redundant terms are included for the sake of the symmetry. For example, the expression to be used for K_B in this case is $K_B = A + C\bar{D}$, where $C\bar{D}$ is redundant, but it makes $J_B = K_B$. Similarly, J_C is made equal to K_C by using an extra term $\bar{A}\bar{B}C\bar{D}$, giving $J_C = K_C = AB + \bar{A}\bar{B}C\bar{D}$, or $J_C = K_C = \overline{AB}\,\overline{\bar{A}\bar{B}C\bar{D}}$ for imple-

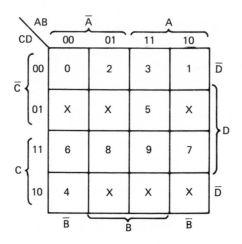

Fig 12.19   Karnaugh map for BCD 2421 Aiken code

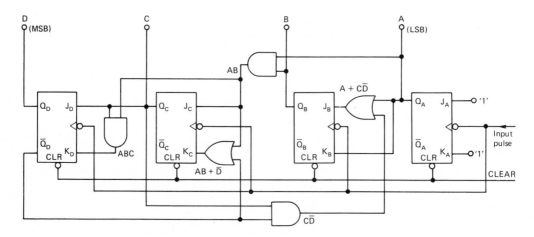

Fig 12.21   BCD 2421 Aiken code counter

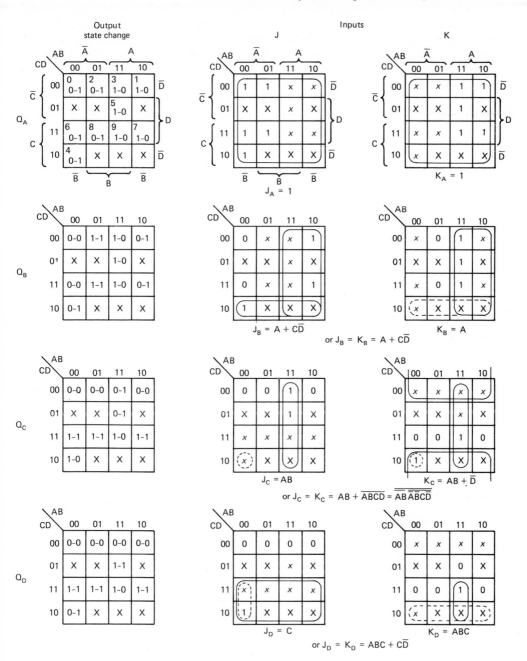

Fig 12.20 Karnaugh maps for BCD 2421 Aiken code counter

mentation with NAND gates as illustrated in Fig. 12.22. The last stage of the counter, however, is implemented with AND gates so that the 7472 IC could be used for this with internally available AND gates. Otherwise one NAND gate with an additional inverter would be required for K_D. $J_D = C$ and therefore it could be fed directly from Q_C. This example illustrates the use of the most convenient gates available for the implementation of the counter rather than satisfying the simplified expressions as derived directly from the Karnaugh maps.

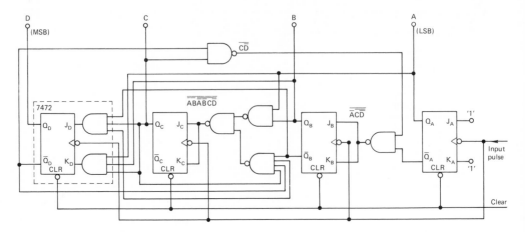

Fig 12.22   Alternative implementation of BCD 2421 Aiken code counter

The BCD 2421 Aiken counter is very useful because the nine's complement of the count number is always available from the $\bar{Q}$ outputs. This facility is used in subtraction since the nine's complement of a number n is $(9 - n)$; for instance, the nine's complement of 7 is 2, etc.

Excess-3 decade counter

The excess-3 decade counter is a BCD 8421 counter but with the first three counts made redundant as shown in the truth table (Table 12.13). The counter is initially preset to start counting at binary 0011 and, for a decade counter, is made to countinue counting to binary

Table 12.13   Transition table for excess-3 counter

Input pulses	Present state Q_n				Next state Q_{n+1}			
	D	C	B	A	D	C	B	A
0	0	0	1	1	0	1	0	0
1	0	1	0	0	0	1	0	1
2	0	1	0	1	0	1	1	0
3	0	1	1	0	0	1	1	1
4	0	1	1	1	1	0	0	0
5	1	0	0	0	1	0	0	1
6	1	0	0	1	1	0	1	0
7	1	0	1	0	1	0	1	1
8	1	0	1	1	1	1	0	0
9	1	1	0	0	0	0	1	1

1100. The six binary numbers 0000, 0001, 0010, 1101, 1110 and 1111 are then the redundant numbers indicated by X's on the Karnaugh maps in Fig. 12.23 and 12.24. The variable A is the least significant bit. Once again, assuming that J-K flip-flops are used, the Karnaugh maps for the interfacing logic are shown in Fig. 12.24.

The excess-3 BCD counter is implemented using J-K flip-flops and NAND gates as shown in Fig. 12.25.

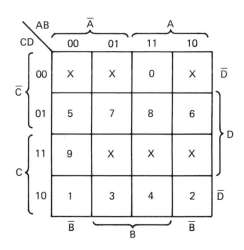

Fig 12.23 Karnaugh map for BCD excess-3 counter

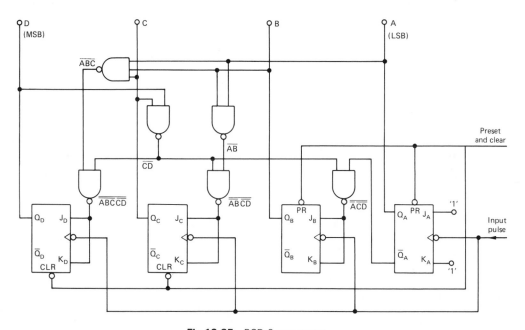

Fig 12.25 BCD-3 up counter

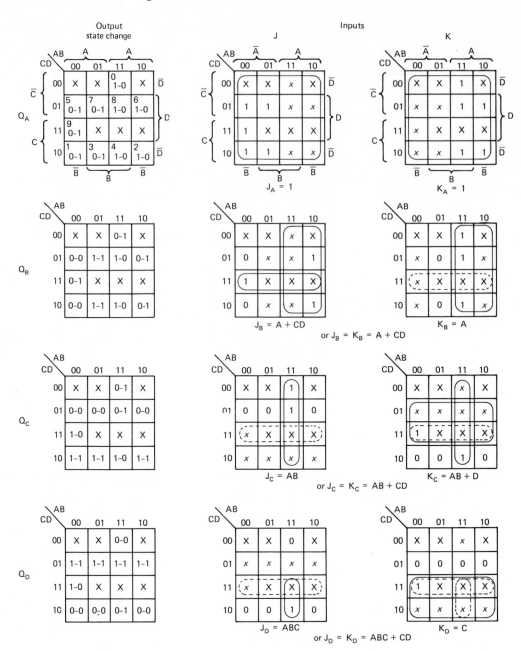

Fig 12.24   Karnaugh maps for BCD excess-3 counter

Gray code decade counters

Gray codes are very useful for numerical machine control and other industrial applications since only one digit changes between two adjacent numbers, which greatly reduces a possibility of error. Two Gray code counters will be considered: reflected binary decade and BCD 2421 Gray code.

Reflected binary Gray code decade counter

The reflected binary Gray code, with its transition table, for a decade counter, is given in Table 12.14.

Table 12.14   Transition table for reflected binary Gray code decade counter

Input pulses	Present state Q_n				Sequence	Next state Q_{n+1}			
	D	C	B	A		D	C	B	A
0	0	0	0	0	0	0	0	0	1
1	0	0	0	1	1	0	0	1	1
2	0	0	1	1	3	0	0	1	0
3	0	0	1	0	2	0	1	1	0
4	0	1	1	0	6	0	1	1	1
5	0	1	1	1	7	0	1	0	1
6	0	1	0	1	5	0	1	0	0
7	0	1	0	0	4	1	1	0	0
8	1	1	0	0	12	1	1	0	1
9	1	1	0	1	13	0	0	0	0

As in the case of the previously described decade codes, the BCD Gray code has also six redundant terms, which are 1000, 1001, 1010, 1011, 1110 and 1111. The six terms are marked with X's in the Karnaugh maps of Figs. 12.26 and 12.27.

Assuming that J-K flip-flops are used, with A as the least significant digit, the design of the counter progresses with preparation of the Karnaugh maps for the output state change of the flip-flops and the maps for the J-K input logic as shown in Fig. 12.26.

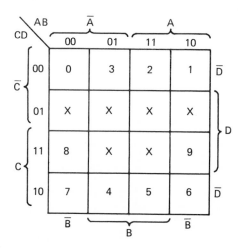

Fig 12.26   Karnaugh map for reflected binary Gray code counter

The counter is implemented in Fig. 12.28 using the simplified expressions from Fig. 12.27. The solution for symmetrical J-K inputs are given in Fig. 12.27 for reference only.

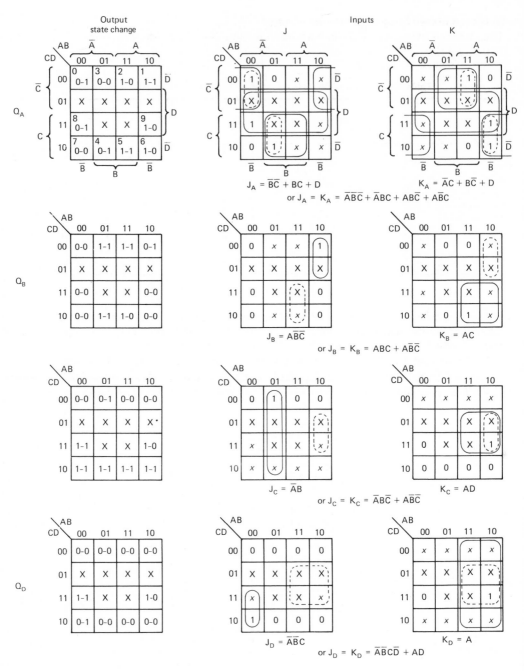

Fig 12.27   Karnaugh maps for reflected binary decade gray code counter

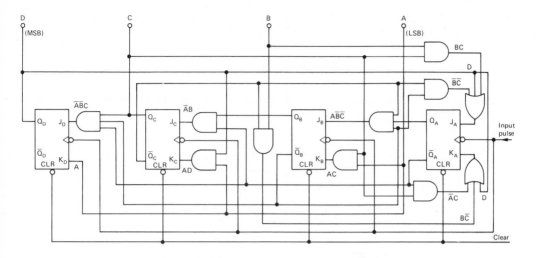

Fig 12.28 Reflected binary Gray code counter

If this counter is implemented with 7472 AND gated flip-flops for B, C and D, then no external gates would be required for these three stages. Only the flip-flop A would need external gates to implement $J_A = \overline{B}\overline{C} + BC + D$ and $K_A - \overline{A}C + B\overline{C} + D$. Alternatively the entire counter could be implemented, if preferred.

BCD 2421 Gray code counter

This is another useful Gray code counter. The truth table for the code together with the output transition table are shown in Table 12.15, and the Karnaugh map for the BCD 2421 Gray code sequence is shown in Fig. 12.29

Table 12.15 Transition table for BCD Gray code

Input pulses	Present state Q_n				Sequence	Next state Q_{n+1}			
	D	C	B	A		D	C	B	A
	2	4	2	1		2	4	2	1
0	0	0	0	0	0	0	0	0	1
1	0	0	0	1	1	0	0	1	1
2	0	0	1	1	3	0	0	1	0
3	0	0	1	0	2	0	1	1	0
4	0	1	1	0	6	1	1	1	0
5	1	1	1	0	14	1	0	1	0
6	1	0	1	0	10	1	0	1	1
7	1	0	1	1	11	1	0	0	1
8	1	0	0	1	9	1	0	0	0
9	1	0	0	0	8	0	0	0	0

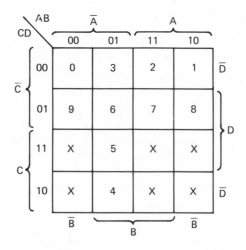

Fig 12.29   Karnaugh map for BCD 2421 Gray code sequence

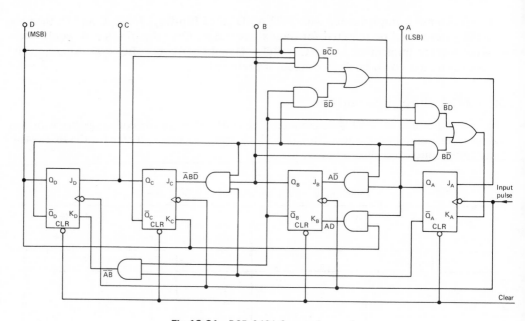

Fig 12.31   BCD 2421 Gray code counter

The design follows the usual pattern as displayed by the Karnaugh maps in Fig. 12.30. The counter is implemented (Fig. 12.31) using the simplified expressions from the Karnaugh maps of Fig. 12.30. Symmetrical solutions for the J and K inputs are also derived, but they need too many gates to be of any practical value.

Note that, although the column weighting of the code is 2421, the sequence values given in Table 12.15 are decimal, pure binary equivalents, of the counter outputs Q_A, Q_B, Q_C and Q_D, with A being the least significant digit.

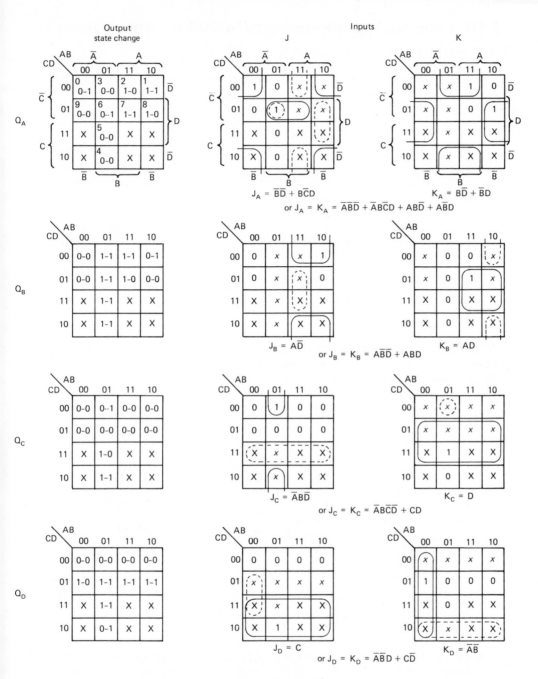

Fig 12.30   Karnaugh maps for BCD 2421 Gray code counter $\overline{A}BD$

12.10 Coupling between decades in BCD counting systems

A brief account on coupling between decades is given here.

When more than one decade is used some form of coupling is necessary, although very often a direct connection from the output of one decade to the input of the next decade is all that is required.

Coupling for asynchronous counters

In asynchronous counters, the coupling between stages is very primitive. The output of one decade is simply coupled to the input of the next decade, as shown in Fig. 12.32 for a ripple through BCD 8421 counter, where A, B, C and D are the Q_A, Q_B, Q_C and Q_D outputs of the flip-flops within each decade respectively.

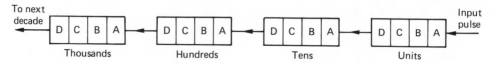

Fig 12.32   Coupling decades of a ripple through counter

Since delays through the stages may give rise to ambiguities, it is necessary to ensure that the propagation delay per stage is not excessive.

When multi stage asynchronous counters operate in the reversible, up/down, mode, the coupling between the stages needs additional logic. One form of coupling using two AND gates and an OR gate, per stage, is shown in Fig. 12.33.

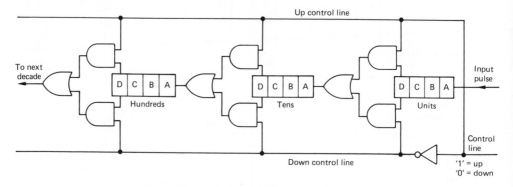

Fig 12.33   AND/OR coupling between decades

Often it may be preferable to use NAND gates. A three NAND gate version of coupling circuit is shown in Fig. 12.34, but this time only two decades are shown. The coupling between additional decades is identical.

Coupling for synchronous counters

Coupling between stages of successive decades in synchronous counters requires more complex logic than that used for asynchronous counters. As an example, three decades of synchronous BCD 8421 counter are shown in Fig. 12.35.

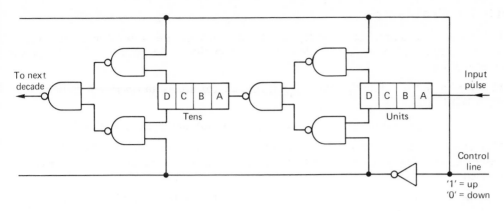

Fig 12.34 NAND gate coupling between decades

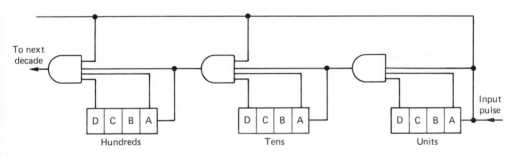

Fig 12.35 Coupling between decades of synchronous BCD 8421 counter

As seen from Fig. 12.35, the transfer of a digit from one decade to the next decade occures only when A, D and the input pulse (the clock) are all at logic 1 level. The logic circuit for coupling decades of up/down (reversible) counter is more complex. An example using NAND gates for coupling is shown in Fig. 12.36.

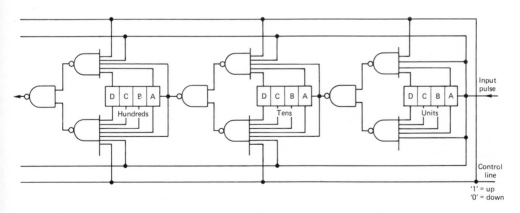

Fig 12.36 Coupling between decades of synchronous up/down counter

12.10 Problems

12.1 Give an example of a code that fulfils each of the following requirements: (a) simplicity; (b) easy error detection; (c) minimal hardware; (d) low power consumption.

12.2 Design a 3 to 8 line decoding circuit suitable for use as an operation code decoder using NOR gates only.

12.3 Design an encoding circuit to convert the hexadecimal values 0 to F to binary.

12.4 Design an alphanumeric code to include upper case letters and 38 special characters. Tabulate the code giving hexadecimal and octal equivalents, assuming that odd parity is to be used.

12.5 Design a circuit to convert the Aiken (2421) BCD code into XS3 BCD code.

12.6 Tabulate a 5 bit Gray code and explain how it is generated.

12.7 Design a synchronous counter to generate the 7421 BCD code sequence given in Table 12.8.

12.8 Design a synchronous counter to generate the 5421 BCD code sequence given in Fig. 12.9.

12.9 Explain how BCD counters can be linked together to provide indications of Units, Tens, Hundreds, Thousands, etc., for indicators such as petrol pump displays.

Appendix A

DESIGN EXERCISES

1 Design a circuit to perform the following functions.

(a) Transfer a 7-bit value from register A to an 8-bit register B. Use parallel transmission.
(b) Generate an odd parity bit.
(c) Transmit the seven bits plus parity bit serially to register C.
(d) Detect whether a single-bit error has occurred.

Construct the circuit on a breadboard and check its operation.

2 Design and construct a logic circuit to add or subtract two 8-bit binary numbers to the following specification.

(a) The circuit is to provide the sum or difference of two signed 8-bit numbers.
(b) If a carry out occurs, a carry flag is to be set.
(c) A 'half carry' flag is to be provided which will be set if a carry occurs from bit 3 to bit 4 to enable BCD arithmetic to be implemented.
(d) A flag to indicate two's complement overflow is also to be provided.
(e) The function of addition or subtraction is to be selected by using a control line CL, so that if CL = 0 the *add* function is selected. If CL = 1 the *subtract* function is implemented.
(f) Parallel operation is to be used throughout.
(g) Provision is to be made for expansion to 16-bit operation.

3 Design an asynchronous counter to follow the sequence

   0, 1, 2, 3, 4, 5, 13, 14, 15

Construct this using JK master–slave flip-flops.

4 Design a synchronous counter to follow the sequence

   0, 13, 1, 2, 5, 4, 3, 9, 14

Use JK flip-flops to construct the circuit.

5 Design an asynchronous modulo-6 counter using SR flip-flops.

6 Design and construct a multiplexer to transmit the contents of an 8-bit register serially to another register without disturbing the original register content.

7 Design and construct an 8-bit hardware timer which loads a timer register with a starting value, decrements the register content to zero, sets a flag, resets the original counter value and stops.

8 Design and construct an 8-bit register system with the following facilities:

(a) Parallel input/output.
(b) Serial input/output.
(c) Right or left shift.
(d) Right or left recirculate.

The function to be carried out is selected by means of a four-bit operation code.

9 Design a circuit to select one pulse in N from a clock signal. N is obtained from a four-bit register content.

10 Design a circuit to generate a three-phase output from an input clock signal. Three square waves, 120° apart in phase, are to be produced.

Appendix B

ADVANCED MINIMISATION TECHNIQUES

B.1 Introduction

Chapter 5 deals with minimisation techniques that cater for functions with up to four variables. As the number of variables increases so does the complexity of the minimisation process. In order to illustrate this, examples follow which demonstrate the minimisation of five and six variable functions using Karnaugh map and tabular methods.

B.2 Five-variable Karnaugh maps

There are two variants of Karnaugh map solutions for five variables:

(i) a map within a map,
(ii) using one bigger Karnaugh map with 32 cells.

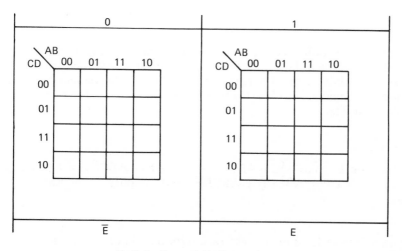

Fig B 1   Five variable Karnaugh map

Both methods will be considered. The first, a map within a map, is shown in Fig. B.1. One four-variable map is used for the variable $\bar{E}$ and the other four-variable map is used for E. The terms which occur in the same position on different maps will be independent of the variable E, and will simplify according to the rules for the four-variable maps.

Alternatively a five-variable Karnaugh map, as shown in Fig. B.2, could be used. Here the variables are divided into two groups. Three variables are designated for the columns and are marked on the right hand side of the diagonal, and the remaining two variables are marked on the left hand side of the diagonal; this is similar to the four-variable maps. For completeness, however, Veitch diagram labelling is also shown, together with the appropriate cell numbering.

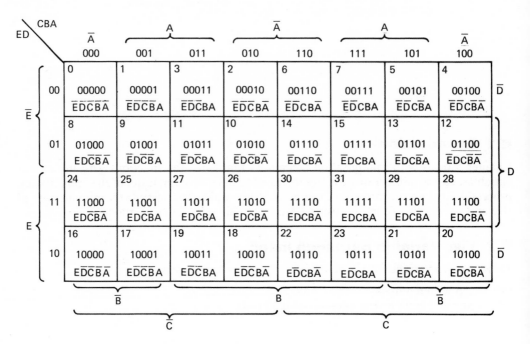

Fig B 2   Alternative five-variable Karnaugh map

Marking the five-variable Karnaugh map with 1's and 0's along the top may cause some problems. A suggestion of how to tackle this is to label the first four columns in the same way as the four-variable maps. Then repeat the labelling for the remaining four columns in the reverse order. Finally, add 0's to the first four columns marked and 1's to the remaining four, as shown in Fig. B.2.

Example B.1

Plot the function $F = \bar{A}\bar{B}CD\bar{E} + \bar{A}\bar{B}CD\bar{E} + \bar{A}\bar{B}CD\bar{E} + ABC\bar{D}E + AB\bar{C}DE + \bar{A}B\bar{C}DE + \bar{A}B\bar{C}D\bar{E} + ABCDE + AB\bar{C}D\bar{E} + \bar{A}B\bar{C}D\bar{E}$, and simplify using Karnaugh mapping techniques.

Solution
(i) Using two four-variable maps, plot each term on one of the two Karnaugh maps as shown in Fig. B.3.

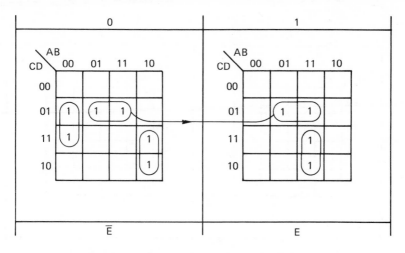

Fig B 3 Karnaugh map for Solution (i) to Example 5.8

The terms ending with $\overline{E}$ are mapped on the left and the terms ending with E are entered on the right hand side of the two four-variable squares. Then, normal looping takes place, allowing for common loops on both maps. Finally, demapping follows, and the simplified solution is produced by writing each term. This gives $F = B\overline{C}D + \overline{A}B\overline{D}\overline{E} + \overline{A}B\overline{C}\overline{E} + ABCE$.

(ii) Using one five-variable map, the given function is mapped as shown in Fig. B.4. Looping and demapping gives $F = B\overline{C}D + \overline{A}B\overline{D}\overline{E} + \overline{A}B\overline{C}\overline{E} + ABCE$, which of course is the same as the solution in (i) above.

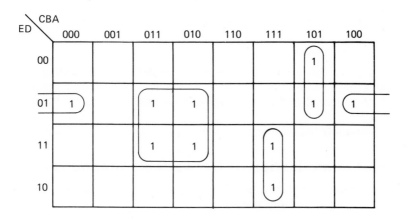

Fig B 4 Five-variable Karnaugh map for Example 5.8

B.3 Six-variable Karnaugh maps

Two solutions are again considered: one using four four-variable maps, as shown in Fig. B.5, and the other using the composite six-variable Karnaugh map of Fig. B.6. The

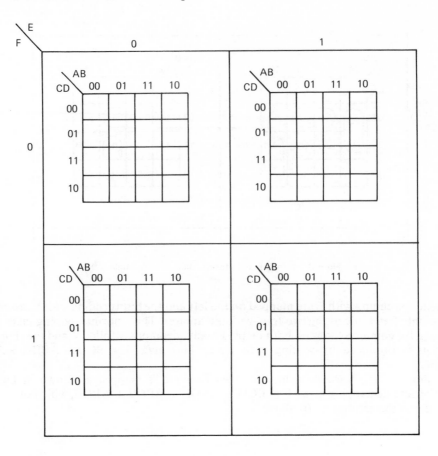

Fig B 5 Four four-variable Karnaugh maps forming six-variable map

map in Fig. B.5 is a further extension of the five-variable technique, adding two extra four-variable squares for the additional variable F. In Fig. B.6 the variable F is added on the left hand side of the diagonal and the number of rows are extended to allow for the sixth variable.

The same rules apply for mapping and demapping as previously discussed.

B.4 Tabular minimisation of five variable functions

The technique of minimising functions of up to four variables in Chapter 5 is now extended to five variables by means of an example.

Example B.2

Simplify the function $F = \overline{V}\,\overline{W}XYZ + VW + WX\overline{Y}Z + XY$, using the tabular method.

Solution

Expanding the function to the canonical form gives:

$$F = \overline{V}\,\overline{W}XYZ + VW(Z + \overline{Z})(XY + X\overline{Y} + \overline{X}Y + \overline{X}\,\overline{Y}) + WX\overline{Y}Z(V + \overline{V})$$
$$+ XY(VW + V\overline{W} + \overline{V}\,W + \overline{V}\,\overline{W})(Z + \overline{Z})$$

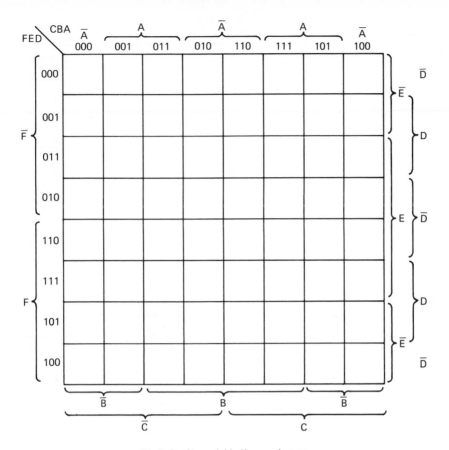

Fig B 6 Six-variable Karnaugh map

$$
\begin{aligned}
&= \overline{V}\,\overline{W}\,XYZ + (VWXY + VWX\overline{Y} + VW\overline{X}Y + VW\overline{X}\,\overline{Y})(Z + \overline{Z}) \\
&\quad + VWX\overline{Y}Z + \overline{V}WX\overline{Y}Z + (VWXY + V\overline{W}XY + \overline{V}WXY \\
&\quad + \overline{V}\,\overline{W}XY)(Z + \overline{Z}) \\
&= \overline{V}\,\overline{W}YZ + VWXYZ + VWXY\overline{Z} + VWX\overline{Y}Z + VWX\overline{Y}\,\overline{Z} \\
&\quad + VW\overline{X}YZ + VW\overline{X}Y\overline{Z} + VW\overline{X}\,\overline{Y}Z + VW\overline{X}\,\overline{Y}\,\overline{Z} + VWX\overline{Y}Z \\
&\quad + \overline{V}WX\overline{Y}Z + VWXYZ + VWXY\overline{Z} + V\overline{W}XYZ + V\overline{W}XY\overline{Z} \\
&\quad + \overline{V}WXYZ + \overline{V}WXY\overline{Z} + \overline{V}\,\overline{W}XYZ + \overline{V}\,\overline{W}XY\overline{Z} \\
&= \overline{V}\,\overline{W}XYZ + VWXYZ + VWXY\overline{Z} + VWX\overline{Y}Z + VWX\overline{Y}\,\overline{Z} \\
&\quad + VW\overline{X}YZ + VW\overline{X}Y\overline{Z} + VW\overline{X}\,\overline{Y}Z + VW\overline{X}\,\overline{Y}\,\overline{Z} + \overline{V}WX\overline{Y}Z \\
&\quad + V\overline{W}XYZ + VWXY\overline{Z} + \overline{V}WXYZ + \overline{V}WXY\overline{Z} + \overline{V}\,\overline{W}XY\overline{Z}
\end{aligned}
$$

The function is now arranged in tabular form, as shown in Tables B.1 to B.4 in lists (a) to (d).

From tables B.3 and B.4 the prime implicants are P, Q, R and S, whcih are represented in a matrix form in Table B.5. By inspection of Table B. 5 it can be seen that only three terms are required. The prime implicant P is covered by S; therefore the minimal form of the function is F = Q + R + S or F = WXZ + VW + XY.

Table B.1

List (a)

No of 1's	Decimal equivalent	Canonical term	Binary equivalent	
2	6	$\bar{V}\bar{W}XY\bar{Z}$	00110	✓
	24	$VW\bar{X}\bar{Y}\bar{Z}$	11000	✓
3	7	$\bar{V}\bar{W}XYZ$	00111	✓
	13	$\bar{V}WX\bar{Y}Z$	01101	✓
	14	$\bar{V}WXY\bar{Z}$	01110	✓
	22	$V\bar{W}XY\bar{Z}$	10110	✓
	25	$VW\bar{X}\bar{Y}Z$	11001	✓
	26	$VW\bar{X}Y\bar{Z}$	11010	✓
	28	$VWX\bar{Y}\bar{Z}$	11100	✓
4	15	$\bar{V}WXYZ$	01111	✓
	23	$V\bar{W}XYZ$	10111	✓
	27	$VW\bar{X}YZ$	11011	✓
	29	$VWX\bar{Y}Z$	11101	✓
	30	$VWXY\bar{Z}$	11110	✓
5	31	$VWXYZ$	11111	✓

Table B.2

List (b)

Decimal equivalent	Binary match	
6,7	0011–	✓
6,14	0–110	✓
6,22	–0110	✓
24,25	1100–	✓
24,26	110–0	✓
24,28	11–00	✓
7,15	0–111	✓
7,23	–0111	✓
13,15	011–1	✓
13,29	–1101	✓
14,15	0111–	✓
14,30	–1110	✓
25,27	110–1	✓
25,29	11–01	✓
26,27	1101–	✓
26,30	11–10	✓
28,29	1110–	✓
28,30	111–0	✓
15,31	–1111	✓
23,31	1–111	✓
27,31	11–11	✓
29,31	111–1	✓
30,31	1111–	✓

Table B.3

List (c)

Decimal equivalent	Binary match	
6,7; 14,15	0–11– } P	
6,14; 7,15	0–11–	
6,22; 7,23	–011–	✓
6,22; 14,30	––110	✓
24,25; 26,27	110––	✓
24,25; 28,29	11–0–	✓
24,26; 25,27	110––	✓
24,26; 28,30	11––0	✓
24,28; 25,29	11–0–	✓
24,28; 26,30	11––0	✓
7,15; 23,31	––111 }	✓
7,23; 15,31	––111	
13,15; 29,31	–11–1 } Q	
13,29; 15,31	–11–1	
14,15; 30,31	–111– }	
14,30; 15,31	–111–	
25,27; 29,31	11––1 }	
25,29; 27,31	11––1	
26,27; 30,31	11–1– }	
26,30; 27,31	11–1–	
28,29; 30,31	111–– }	
28,30; 29,31	111––	

Table B.4

List (d)

Decimal equivalent	Binary match	
6,22; 7,23/14,15; 30,31	––11– } R	
6.22; 14,30/7,15; 23,31	––11–	
24,25; 26,27/28,29; 30,31	11––– }	
24,25; 28,29/26,27; 30,31	11––– } S	
24,26; 28,30/25,27; 29,31	11–––	

Table B.5   Matrix of prime implicants

	6	7	13	14	15	22	23	24	25	26	27	28	29	30	31
P: 6,7,14,15	X	X		X	X										
Q: 13,15,29,31			X		X								X		X
R: 24,25,26,27,28,29,30,31								X	X	X	X	X	X	X	X
S: 6,7,14,15,22,23,30,31	X	X		X	X	X	X							X	X

Table B.6

Decimal value	No. of 1's	A	B	C	X	Y	Z
4	1	0	0	0	1	0	0
6	2	0	0	0	1	1	0
7	3	0	0	0	1	1	1
8	1	0	0	1	0	0	0
9	2	0	0	1	0	0	1
10	2	0	0	1	0	1	0
11	3	0	0	1	0	1	1
12	2	0	0	1	1	0	0
13	3	0	0	1	1	0	1
14	3	0	0	1	1	1	0
15	4	0	0	1	1	1	1
32	1	1	0	0	0	0	0
33	2	1	0	0	0	0	1
34	2	1	0	0	0	1	0
35	3	1	0	0	0	1	1
36	2	1	0	0	1	0	0
37	3	1	0	0	1	0	1
38	3	1	0	0	1	1	0
39	4	1	0	0	1	1	1
40	2	1	0	1	0	0	0
41	3	1	0	1	0	0	1
42	3	1	0	1	0	1	0
43	4	1	0	1	0	1	1
44	3	1	0	1	1	0	0
45	4	1	0	1	1	0	1
46	4	1	0	1	1	1	0
47	5	1	0	1	1	1	1
48	2	1	1	0	0	0	0
49	3	1	1	0	0	0	1
50	3	1	1	0	0	1	0
51	4	1	1	0	0	1	1
52	3	1	1	0	1	0	0
53	4	1	1	0	1	0	1
54	4	1	1	0	1	1	0
55	5	1	1	0	1	1	1
56	3	1	1	1	0	0	0
57	4	1	1	1	0	0	1
58	4	1	1	1	0	1	0
59	5	1	1	1	0	1	1
60	4	1	1	1	1	0	0
61	5	1	1	1	1	0	1
62	5	1	1	1	1	1	0
63	6	1	1	1	1	1	1

Table B.7   Tabular minimisation. Lists (a) to (f)

List (a)

	A	B	C	X	Y	Z	
4	0	0	0	1	0	0	✓
8	0	0	1	0	0	0	✓
32	1	0	0	0	0	0	✓
6	0	0	0	1	1	0	✓
9	0	0	1	0	0	1	✓
10	0	0	1	0	1	0	✓
12	0	0	1	1	0	0	✓
33	1	0	0	0	0	1	✓
34	1	0	0	0	1	0	✓
36	1	0	0	1	0	0	✓
40	1	0	1	0	0	0	✓
48	1	1	0	0	0	0	✓
7	0	0	0	1	1	1	✓
11	0	0	1	0	1	1	✓
13	0	0	1	1	0	1	✓
14	0	0	1	1	1	0	✓
35	1	0	0	0	1	1	✓
37	1	0	0	1	0	1	✓
38	1	0	0	1	1	0	✓
41	1	0	1	0	0	1	✓
42	1	0	1	0	1	0	✓
44	1	0	1	1	0	0	✓
49	1	1	0	0	0	1	✓
50	1	1	0	0	1	0	✓
52	1	1	0	1	0	0	✓
56	1	1	1	0	0	0	✓
15	0	0	1	1	1	1	✓
39	1	0	0	1	1	1	✓
43	1	0	1	0	1	1	✓
45	1	0	1	1	0	1	✓
46	1	0	1	1	1	0	✓
51	1	1	0	0	1	1	✓
53	1	1	0	1	0	1	✓
54	1	1	0	1	1	0	✓
57	1	1	1	0	0	1	✓
58	1	1	1	0	1	0	✓
60	1	1	1	1	0	0	✓
47	1	0	1	1	1	1	✓
55	1	1	0	1	1	1	✓
59	1	1	1	0	1	1	✓
61	1	1	1	1	0	1	✓
62	1	1	1	1	1	0	✓
63	1	1	1	1	1	1	✓

List (b)

	A	B	C	X	Y	Z	
4, 6	0	0	0	1	–	0	✓
4,12	0	0	–	1	0	0	✓
4,36	–	0	0	1	0	0	✓
8, 9	0	0	1	0	0	–	✓
8,10	0	0	1	0	–	0	✓
8,40	–	0	1	0	0	0	✓

List (b) *cont.*

	A	B	C	X	Y	Z	
32,33	1	0	0	0	0	–	✓
32,34	1	0	0	0	–	0	✓
32,36	1	0	0	–	0	0	✓
32,40	1	0	–	0	0	0	✓
32,48	1	–	0	0	0	0	✓
6, 7	0	0	0	1	1	–	✓
6,14	0	0	–	1	1	0	✓
6,38	–	0	0	1	1	0	✓
9,11	0	0	1	0	–	1	✓
9,13	0	0	1	–	0	1	✓
9,41	–	0	1	0	0	1	✓
10,11	0	0	1	0	1	–	✓
10,14	0	0	1	–	1	0	✓
10,42	–	0	1	0	1	0	✓
12,13	0	0	1	1	0	–	✓
12,14	0	0	1	1	–	0	✓
12,44	–	0	1	1	0	0	✓
33,35	1	0	0	0	–	1	✓
33,37	1	0	0	–	0	1	✓
33,41	1	0	–	0	0	1	✓
33,49	1	–	0	0	0	1	✓
34,35	1	0	0	0	1	–	✓
34,38	1	0	0	–	1	0	✓
34,42	1	0	–	0	1	0	✓
34,50	1	–	0	0	1	0	✓
36,37	1	0	0	1	0	–	✓
36,38	1	0	0	1	–	0	✓
36,44	1	0	–	1	0	0	✓
36,52	1	–	0	1	0	0	✓
40,41	1	0	1	0	0	–	✓
40,42	1	0	1	0	–	0	✓
40,44	1	0	1	–	0	0	✓
40,56	1	–	1	0	0	0	✓
48,49	1	1	0	0	0	–	✓
48,50	1	1	0	0	–	0	✓
48,52	1	1	0	–	0	0	✓
48,56	1	1	–	0	0	0	✓
7,15	0	0	–	1	1	1	✓
7,39	–	0	0	1	1	1	✓
11,15	0	0	1	–	1	1	✓
11,43	–	0	1	0	1	1	✓
13,15	0	0	1	1	–	1	✓
13,45	–	0	1	1	0	1	✓
14,15	0	0	1	1	1	–	✓
14,46	–	0	1	1	1	0	✓
35,39	1	0	0	–	1	1	✓
35,43	1	0	–	0	1	1	✓
35,51	1	–	0	0	1	1	✓
37,39	1	0	0	1	–	1	✓
37,45	1	0	–	1	0	1	✓
37,53	1	–	0	1	0	1	✓
38,39	1	0	0	1	1	–	✓
38,46	1	0	–	1	1	0	✓
38,54	1	–	0	1	1	0	✓
41,43	1	0	1	0	–	1	✓
41,45	1	0	1	–	0	1	✓
41,57	1	–	1	0	0	1	✓

List (b) *cont.*

	A	B	C	X	Y	Z	
42,43	1	0	1	0	1	–	✓
42,46	1	0	1	–	1	0	✓
42,58	1	–	1	0	1	0	✓
44,45	1	0	1	1	0	–	✓
44,46	1	0	1	1	–	0	✓
44,60	1	–	1	1	0	0	✓
49,51	1	1	0	0	–	1	✓
49,53	1	1	0	–	0	1	✓
49,57	1	1	–	0	0	1	✓
50,51	1	1	0	0	1	–	✓
50,54	1	1	0	–	1	0	✓
50,58	1	1	–	0	1	0	✓
52,53	1	1	0	1	0	–	✓
52,54	1	1	0	1	–	0	✓
52,60	1	1	–	1	0	0	✓
56,57	1	1	1	0	0	–	✓
56,58	1	1	1	0	–	0	✓
56,60	1	1	1	–	0	0	✓
15,47	–	0	1	1	1	1	✓
39,47	1	0	–	1	1	1	✓
39,55	1	–	0	1	1	1	✓
43,47	1	0	1	–	1	1	✓
43,59	1	–	1	0	1	1	✓
45,47	1	0	1	1	–	1	✓
45,61	1	–	1	1	0	1	✓
46,47	1	0	1	1	1	–	✓
46,62	1	–	1	1	1	0	✓
51,55	1	1	0	–	1	1	✓
51,59	1	1	–	0	1	1	✓
53,55	1	1	0	1	–	1	✓
53,61	1	1	–	1	0	1	✓
54,55	1	1	0	1	1	–	✓
54,62	1	1	–	1	1	0	✓
57,59	1	1	1	0	–	1	✓
57,61	1	1	1	–	0	1	✓
58,59	1	1	1	0	1	–	✓
58,62	1	1	1	–	1	0	✓
60,61	1	1	1	1	0	–	✓
60,62	1	1	1	1	–	0	✓
47,63	1	–	1	1	1	1	✓
55,63	1	1	–	1	1	1	✓
59,63	1	1	1	–	1	1	✓
61,63	1	1	1	1	–	1	✓
62,63	1	1	1	1	1	–	✓

List (c)

	A	B	C	X	Y	Z	
4, 6; 12,14	0	0	–	1	–	0	✓
4, 6; 36,38	–	0	0	1	–	0	✓
4,12; 6,14	0	0	–	1	–	0	✓
4,12; 36,44	–	0	–	1	0	0	✓
4,36; 6,39	–	0	0	1	–	0	✓
4,36; 12,44	–	0	–	1	0	0	✓
8, 9; 10,11	0	0	1	0	–	–	✓
8, 9; 12,13	0	0	1	–	0	–	✓

Table B.7   (continued)

List (c)   cont.

	A	B	C	X	Y	Z	
8, 9; 40,41	–	0	1	0	0	–	✓
8,10; 9,11	0	0	1	0	–	–	✓
8,10; 12,14	0	0	1	–	–	0	✓
8,10; 40,42	–	0	1	0	–	0	✓
8,40; 9,41	–	0	1	0	0	–	✓
8,40; 10,42	–	0	1	0	–	0	✓
8,40; 12,44	–	0	1	–	0	0	✓
32,33; 34,35	1	0	0	0	–	–	✓
32,33; 36,37	1	0	0	–	0	–	✓
32,33; 40,41	1	0	–	0	0	–	✓
32,33; 48,49	1	–	0	0	0	–	✓
32,34; 33,35	1	0	0	0	–	–	✓
32,34; 36,38	1	0	0	–	–	0	✓
32,34; 40,42	1	0	–	0	–	0	✓
32,34; 48,50	1	–	0	0	–	0	✓
32,36; 33,37	1	0	0	–	0	–	✓
32,36; 34,38	1	0	0	–	–	0	✓
32,36; 40,44	1	0	–	–	0	0	✓
32,36; 48,52	1	–	0	–	0	0	✓
32,40; 33,41	1	0	–	0	0	–	✓
32,40; 34,42	1	0	–	0	–	0	✓
32,40; 36,44	1	0	–	–	0	0	✓
32,40; 48,56	1	–	–	0	0	0	✓
32,48; 33,49	1	–	0	0	0	–	✓
32,48; 34,50	1	–	0	0	–	0	✓
32,48; 36,52	1	–	0	–	0	0	✓
32,48; 40,56	1	–	–	0	0	0	✓
6, 7; 14,15	0	0	–	1	1	–	✓
6, 7; 38,39	–	0	0	1	1	–	✓
6,14; 7,15	0	0	–	1	1	–	✓
6,14; 38,46	–	0	–	1	1	0	✓
6,38; 7,39	–	0	0	1	1	–	✓
6,38; 14,46	–	0	–	1	1	0	✓
9,11; 13,15	0	0	1	–	–	1	✓
9,11; 41,43	–	0	1	0	–	1	✓
9,13; 11,15	0	0	1	–	–	1	✓
9,13; 41,45	–	0	1	–	0	1	✓
9,41; 11,43	–	0	1	0	–	1	✓
9,41; 13,45	–	0	1	–	0	1	✓
10,11; 14,15	0	0	1	–	1	–	✓
10,11; 42,43	–	0	1	0	1	–	✓
10,14; 11,15	0	0	1	–	1	–	✓
10,14; 42,56	–	0	1	–	1	0	✓
10,42; 11,43	–	0	1	0	1	–	✓
10,42; 14,46	–	0	1	–	1	0	✓
12,13; 14,15	0	0	1	1	–	–	✓
12,13; 44,45	–	0	1	1	0	–	✓
12,14; 13,15	0	0	1	1	–	–	✓
12,44; 14,46	–	0	1	1	–	0	✓
12,44; 13,45	–	0	1	1	0	–	✓
12,44; 14,46	–	0	1	1	–	0	✓
33,35; 37,39	1	0	0	–	–	1	✓
33,35; 41,43	1	0	–	0	–	1	✓
33,35; 49,51	1	–	0	0	–	1	✓
33,37; 35,39	1	0	0	–	–	1	✓
33,37; 41,45	1	0	–	–	0	1	✓
33,37; 49,53	1	–	0	–	0	1	✓

List (c)   cont.

	A	B	C	X	Y	Z	
33,41; 35,43	1	0	–	0	–	1	✓
33,41; 49,57	1	–	–	0	0	1	✓
33,49; 35,51	1	–	0	0	–	1	✓
33,49; 37,53	1	–	0	–	0	1	✓
33,49; 41,57	1	–	–	0	0	1	✓
34,35; 38,39	1	0	0	–	1	–	✓
34,35; 42,43	1	0	–	0	1	–	✓
34,35; 50,51	1	–	0	0	1	–	✓
34,38; 35,39	1	0	0	–	1	–	✓
34,38; 42,46	1	0	–	–	1	0	✓
34,38; 50,54	1	–	0	–	1	0	✓
34,42; 35,43	1	0	–	0	1	–	✓
34,42; 38,46	1	0	–	–	1	0	✓
34,42; 50,58	1	–	–	0	1	0	✓
34,50; 35,51	1	–	0	0	1	–	✓
34,50; 38,54	1	–	0	–	1	0	✓
34,50; 42,58	1	–	–	0	1	0	✓
36,37; 38,39	1	0	0	1	–	–	✓
36,37; 44,45	1	0	–	1	0	–	✓
36,37; 52,53	1	–	0	1	0	–	✓
36,38; 37,39	1	0	0	1	–	–	✓
36,38; 44,46	1	–	0	1	–	0	✓
36,38; 52,54	1	–	0	1	–	0	✓
36,44; 37,45	1	0	–	1	0	–	✓
36,44; 38,46	1	0	–	1	–	0	✓
36,44; 52,60	1	–	–	1	0	0	✓
36,52; 37,53	1	–	0	1	0	–	✓
36,52; 38,54	1	–	0	1	–	0	✓
36,52; 44,60	1	–	–	1	0	0	✓
40,41; 42,43	1	0	1	0	–	–	✓
40,41; 44,45	1	0	1	–	0	–	✓
40,41; 56,57	1	–	1	0	0	–	✓
40,42; 41,43	1	0	1	0	–	–	✓
40,42; 44,46	1	0	1	–	–	0	✓
40,42; 56,58	1	–	1	0	–	0	✓
40,44; 42,46	1	0	1	–	–	0	✓
40,44; 56,60	1	–	1	–	0	0	✓
40,56; 41,57	1	–	1	0	0	–	✓
40,56; 42,58	1	–	1	0	–	0	✓
40,56; 44,60	1	–	1	–	0	0	✓
48,49; 50,51	1	1	0	0	–	–	✓
48,49; 52,53	1	1	0	–	0	–	✓
48,49; 56,57	1	1	–	0	0	–	✓
48,50; 49,51	1	1	0	0	–	–	✓
48,50; 52,54	1	1	0	–	–	0	✓
48,50; 56,58	1	1	–	0	–	0	✓
48,52; 49,53	1	1	0	–	0	–	✓
48,52; 50,54	1	1	0	–	–	0	✓
48,52; 56,60	1	1	–	–	0	0	✓
48,56; 49,57	1	1	–	0	0	–	✓
48,56; 50,58	1	1	–	0	–	0	✓
48,56; 52,60	1	1	–	–	0	0	✓
7,15; 39,47	–	0	–	1	1	1	✓
7,39; 15,47	–	0	–	1	1	1	✓
11,15; 43,47	–	0	1	–	1	1	✓
11,43; 45,47	–	0	1	–	1	1	✓
13,15; 45,47	–	0	1	1	–	1	✓

Table B.7   *(continued)*

List (c)   *cont.*

	A	B	C	X	Y	Z	
13,45; 15,47	–	0	1	1	–	1	✓
14,15; 46,47	–	0	1	1	1	–	✓
14,46; 15,47	–	0	1	1	1	–	✓
35,39; 43,47	1	0	–	–	1	1	✓
35,43; 39,47	1	0	–	–	1	1	✓
35,51; 39,51	1	–	0	–	1	1	✓
35,51; 43,59	1	–	–	0	1	1	✓
37,39; 45,47	1	0	–	1	–	1	✓
37,39; 53,55	1	–	0	1	–	1	✓
37,45; 39,47	1	0	–	1	–	1	✓
37,53; 45,61	1	–	–	1	0	1	✓
38,39; 46,47	1	0	–	1	1	–	✓
38,39; 54,55	1	–	0	1	1	–	✓
38,46; 39,47	1	0	–	1	1	–	✓
38,46; 54,62	1	–	–	1	1	0	✓
38,54; 39,55	1	–	0	1	1	–	✓
38,54; 46,62	1	–	–	1	1	0	✓
41,43; 45,47	1	0	1	–	–	1	✓
41,43; 57,59	1	–	1	0	–	1	✓
41,45; 43,47	1	0	1	–	–	1	✓
41,57; 43,59	1	–	1	0	–	1	✓
41,57; 45,61	1	–	1	–	0	1	✓
42,43; 46,47	1	0	1	–	1	–	✓
42,43; 58,59	1	–	1	0	1	–	✓
42,46; 43,47	1	0	1	–	1	–	✓
42,46; 58,62	1	–	1	–	1	0	✓
42,58; 43,59	1	–	1	0	1	–	✓
42,58; 46,62	1	–	1	–	1	0	✓
44,45; 46,47	1	0	1	1	–	–	✓
44,45; 60,61	1	–	1	1	0	–	✓
44,46; 45,47	1	0	1	1	–	–	✓
44,60; 45,61	1	–	1	1	0	–	✓
44,60; 46,62	1	–	1	1	–	0	✓
49,51; 53,55	1	1	0	–	–	1	✓
49,53; 51,55	1	1	0	–	–	1	✓
49,53; 57,61	1	1	–	–	0	1	✓
49,57; 51,59	1	1	–	0	–	1	✓
49,57; 53,61	1	1	–	–	0	1	✓
50,51; 54,55	1	1	0	–	1	–	✓
50,51; 58,59	1	1	–	0	1	–	✓
50,54; 51,55	1	1	0	–	1	–	✓
50,54; 58,62	1	1	–	–	1	0	✓
50,58; 51,59	1	1	–	0	1	–	✓
50,58; 54,62	1	1	–	–	1	0	✓
52,53; 54,55	1	1	0	1	–	–	✓
52,53; 60,61	1	1	–	1	0	–	✓
52,54; 53,55	1	1	0	1	–	–	✓
52,54; 60,61	1	1	–	1	–	0	✓
52,60; 53,61	1	1	–	1	0	–	✓
52,60; 54,62	1	1	–	1	–	0	✓
56,57; 58,59	1	1	1	0	–	–	✓
56,57; 60,61	1	1	1	–	0	–	✓
56,58; 57,59	1	1	1	0	–	–	✓
56,58; 60,62	1	1	1	–	–	0	✓
56,60; 57,61	1	1	1	–	0	–	✓
56,60; 58,62	1	1	1	–	–	0	✓
39,47; 55,63	1	–	–	1	1	1	✓

List (c)   *cont.*

	A	B	C	X	Y	Z	
39,55; 47,63	1	–	–	1	1	1	✓
43,47; 59,61	1	–	1	–	1	1	✓
43,59; 47,63	1	–	1	–	1	1	✓
45,47; 61,63	1	–	1	1	–	1	✓
45,61; 47,63	1	–	1	1	–	1	✓
46,47; 62,63	1	–	1	1	1	–	✓
46,62; 47,63	1	–	1	1	1	–	✓
51,55; 59,61	1	1	–	–	1	1	✓
51,59; 55,63	1	1	–	–	1	1	✓
53,55; 61,63	1	1	–	1	–	1	✓
53,61; 55,63	1	1	–	1	–	1	✓
54,55; 62,63	1	1	–	1	1	–	✓
54,62; 55,63	1	1	–	1	1	–	✓
57,59; 61,63	1	1	1	–	–	1	✓
57,61; 59,63	1	1	1	–	–	1	✓
58,59; 62,63	1	1	1	–	1	–	✓
58,62; 59,63	1	1	1	–	1	–	✓
60,61; 62,63	1	1	1	1	–	–	✓
60,62; 61,63	1	1	1	1	–	–	✓

List (d)

	A	B	C	X	Y	Z	
4, 6; 12,14/36,38; 44,46	–	0	–	1	–	0	} P
4, 6; 36,38/12,14; 44,46	–	0	–	1	–	0	
4,12; 36,44/ 6,14; 38,46	–	0	–	1	–	0	
8, 9; 10,11/12,13; 14,15	0	0	1	–	–	–	Q
8, 9; 10,11/40,41; 42,43	–	0	1	0	–	–	✓
8, 9; 12,13/10,11; 14,15	0	0	1	–	–	–	✓
8, 9; 12,13/40,41; 44,45	–	0	1	–	0	–	✓
8, 9; 40,41/10,11; 42,43	–	0	1	0	–	–	✓
8, 9; 40,41/12,13; 44,45	–	0	1	–	0	–	✓
8,10; 12,14/ 9,11; 13,15	0	0	1	–	–	–	✓
8,10; 12,14/40,42; 44,46	–	0	1	–	–	0	✓
8,10; 40,42/ 9,11; 41,43	–	0	1	0	–	–	✓
8,10; 40,42/12,14; 44,46	–	0	1	–	–	0	✓
8,40; 12,44/ 9,13; 41,45	–	0	1	–	0	–	✓
8,40; 12,44/10,14; 42,46	–	0	1	–	–	0	✓
32,33; 34,35/36,37; 38,39	1	0	0	–	–	–	✓
32,33; 34,35/40,41; 42,43	1	0	–	0	–	–	✓
32,33; 34,35/48,49; 50,51	1	–	0	0	–	–	✓
32,33; 36,37/34,35; 38,39	1	0	0	–	–	–	✓
32,33; 36,37/40,42; 44,45	1	0	–	–	0	–	✓
32,33; 36,37/48,49; 52,53	1	–	0	–	0	–	✓
32,33; 40,41/34,35; 42,43	1	0	–	0	–	–	✓
32,33; 40,41/36,37; 44,45	1	0	–	–	0	–	✓
32,33; 40,41/48,49; 56,57	1	–	–	0	0	–	✓
32,33; 48,49/34,35; 50,51	1	–	0	0	–	–	✓
32,33; 48,49/36,37; 52,53	1	–	0	–	0	–	✓
32,33; 48,49/40,41; 56,57	1	–	–	0	0	–	✓
32,34; 33,35/36,37; 38,39	1	0	0	–	–	–	✓
32,34; 33,35/40,41; 42,43	1	0	–	0	–	–	✓
32,34; 33,35/48,49; 50,51	1	–	0	0	–	–	✓
32,34; 36,38/33,35; 37,39	1	0	0	–	–	–	✓
32,34; 36,38/40,42; 44,46	1	0	–	–	–	0	✓
32,34; 36,38/48,50; 52,54	1	.	0	–	–	0	✓
32,34; 40,42/33,35; 41,43	1	0	–	0	–	–	✓

Table B.7 *(continued)*

List (d)   *cont.*

	A	B	C	X	Y	Z	
32,34; 40,42/36,38; 44,46	1	0	–	–	–	0	✓
32,34; 40,42/48,50; 56,58	1	–	–	0	–	0	✓
32,34; 48,50/33,35; 49,51	1	–	0	0	–	–	✓
32,34; 48,50/36,38; 52,54	1	–	0	–	–	0	✓
32,34; 48,50/40,42; 56,58	1	–	–	0	–	0	✓
32,36; 40,44/33,37; 41,45	1	0	–	–	0	–	✓
32,36; 40,44/34,38; 42,46	1	0	–	–	–	0	✓
32,36; 40,44/48,52; 56,60	1	–	–	–	0	0	✓
32,36; 48,52/33,37; 49,53	1	–	0	–	0	–	✓
32,36; 48,52/34,38; 50,54	1	–	0	–	–	0	✓
32,36; 48,52/40,44; 56,60	1	–	–	–	0	0	✓
32,40; 48,56/33,41; 49,57	1	–	–	0	0	–	✓
32,40; 48,56/34,42; 50,58	1	–	–	0	–	0	✓
32,40; 48,56/36,44; 52,60	1	–	–	–	0	0	✓
6, 7; 14,15/38,39; 46,47	–	0	–	1	1	–	R
6, 7; 38,39/14,15; 46,47	–	0	–	1	1	–	✓
6,14; 38,46/ 7,15; 39,47	–	0	–	1	1	–	✓
9,11; 13,15/41,43; 45,47	–	0	1	–	–	1	✓
9,11; 41,43/13,15; 45,47	–	0	1	–	–	1	✓
9,13; 41,45/11,15; 43,47	–	0	1	–	–	1	✓
10,11; 14,15/42,43; 46,47	–	0	1	–	1	–	✓
10,11; 42,43/14,15; 46,47	–	0	1	–	1	–	✓
10,14; 42,46/11,15; 46,47	–	0	1	–	1	–	✓
12,13; 14,15/44,45; 46,47	–	0	1	1	–	–	✓
12,13; 44,45/14,15; 46,47	–	0	1	1	–	–	✓
12,14; 44,46/13,15; 45,47	–	0	1	1	–	–	✓
33,35; 37,39/41,43; 45,47	1	0	–	–	–	1	✓
33,35; 37,39/49,51; 53,55	1	–	0	–	–	1	✓
33,35; 41,43/37,39; 45,47	1	0	–	–	–	1	✓
33,35; 41,43/49,57; 51,59	1	–	–	0	–	1	✓
33,35; 49,51/37,39; 53,55	1	–	0	–	–	1	✓
33,35; 49,51/41,43; 57,59	1	–	–	0	–	1	✓
33,37; 41,45/35,39; 43,47	1	0	–	–	–	1	✓
33,37; 41,45/49,53; 57,61	1	–	–	–	0	1	✓
33,37; 49,53/35,51; 39,55	1	–	0	–	–	1	✓
33,37; 49,53/41,57; 45,61	1	–	–	–	0	1	✓
33,41; 49,57/35,51; 39,55	1	–	–	0	–	1	✓
33,41; 49,57/37,53; 45,61	1	–	–	–	0	1	✓
34,35; 38,39/42,43; 46,47	1	0	–	–	1	–	✓
34,35; 38,39/50,51; 54,55	1	–	0	–	1	–	✓
34,35; 42,43/38,39; 46,47	1	0	–	–	1	–	✓
34,35; 50,51/38,39; 54,55	1	–	0	–	1	–	✓
34,35; 50,51/42,43; 58,59	1	–	–	0	1	–	✓
34,38; 42,46/35,39; 43,47	1	0	–	–	1	–	✓
34,38; 50,54/35,51; 39,55	1	–	0	–	1	–	✓
34,38; 50,54/42,46; 58,62	1	–	–	–	1	0	✓
34,42; 50,58/35,51; 43,59	1	–	–	0	1	–	✓
34,42; 50,58/38,46; 54,62	1	–	–	–	1	0	✓
36,37; 38,39/44,45; 46,47	1	0	–	1	–	–	✓
36,37; 38,39/52,53; 54,55	1	–	0	1	–	–	✓
36,37; 44,45/38,39; 46,47	1	0	–	1	–	–	✓
36,37; 52,53/38,39; 54,55	1	–	0	1	–	–	✓
36,37; 52,53/44,45; 60,61	1	–	–	1	0	–	✓
36,38; 44,46/37,39; 45,47	1	0	–	1	–	–	✓
36,38; 52,54/37,39; 53,55	1	–	0	1	–	–	✓

List (d)   *cont.*

	A	B	C	X	Y	Z	
36,38; 52,54/44,60; 46,62	1	–	–	1	–	0	✓
36,44; 52,60/37,52; 45,61	1	–	–	1	0	–	✓
36,44; 52,60/38,46; 54,62	1	–	–	1	–	0	✓
40,41; 42,43/44,45; 46,47	1	0	1	–	–	–	✓
40,41; 42,43/56,57; 58,59	1	–	1	0	–	–	✓
40,41; 44,45/42,43; 46,47	1	0	1	–	–	–	✓
40,41; 56,57/42,43; 58,59	1	–	1	–	0	–	✓
40,41; 56,47/44,45; 60,61	1	–	1	–	0	–	✓
40,42; 44,46/41,43; 45,47	1	0	1	–	–	–	✓
40,42; 56,58/41,43; 57,59	1	–	1	0	–	–	✓
40,42; 56,58/44,60; 46,62	1	–	1	–	–	0	✓
40,44; 56,60/41,57; 45,61	1	–	1	–	0	–	✓
40,44; 56,60/42,46; 58,62	1	–	1	–	–	0	✓
48,49; 50,51/52,53; 54,55	1	1	0	–	–	–	✓
48,49; 50,51/56,57; 58,59	1	1	–	0	–	–	✓
48,49; 52,53/50,51; 54,55	1	1	0	–	–	–	✓
48,49; 52,53/56,57; 60,61	1	1	–	–	0	–	✓
48,49; 56,57/50,51; 58,59	1	1	–	0	–	–	✓
48,49; 56,57/52,53; 60,61	1	1	–	–	0	–	✓
48,50; 52,54/49,51; 53,55	1	1	0	–	–	–	✓
48,50; 52,54/56,58; 60,62	1	1	–	–	–	0	✓
48,50; 56,58/49,57; 51,59	1	1	–	0	–	–	✓
48,50; 56,58/52,54; 60,62	1	1	–	–	–	0	✓
48,52; 56,60/49,53; 57,61	1	1	–	–	0	–	✓
48,52; 56,60/50,58; 54,62	1	1	–	–	–	0	✓
35,39; 43,47/51,55; 59,63	1	–	–	–	1	1	✓
35,51; 39,55/43,47; 59,63	1	–	–	–	1	1	✓
35,51; 43,59/39,47; 55,63	1	–	–	–	1	1	✓
37,39; 45,47/53,55; 61,63	1	–	–	1	–	1	✓
37,39; 53,55/45,47; 61,63	1	–	–	1	–	1	✓
37,53; 45,61/39,47; 55,63	1	–	–	1	–	1	✓
38,39; 46,47/54,55; 62,63	1	–	–	1	1	–	✓
38,39; 54,55/46,47; 62,63	1	–	–	1	1	–	✓
38,46; 54,62/39,47; 55,63	1	–	–	1	1	–	✓
41,43; 45,47/57,59; 61,63	1	–	1	–	–	1	✓
41,43; 57,59/45,47; 61,63	1	–	1	–	–	1	✓
41,57; 45,61/43,47; 59,63	1	–	1	–	–	1	✓
42,43; 46,47/58,59; 62,63	1	–	1	–	1	–	✓
42,43; 58,59/46,47; 62,63	1	–	1	–	1	–	✓
42,46; 58,62/43,47; 59,63	1	–	1	–	1	–	✓
44,45; 46,47/60,61; 62,63	1	–	1	1	–	–	✓
44,45; 60,61/46,47; 62,63	1	–	1	1	–	–	✓
44,60; 46,62/45,47; 61,63	1	–	1	1	–	–	✓
49,51; 53,55/57,59; 61,63	1	1	–	–	–	1	✓
49,53; 57,61/51,55; 59,63	1	1	–	–	–	1	✓
49,57; 51,59/53,55; 61,63	1	1	–	–	–	1	✓
50,51; 54,55/58,59; 62,63	1	1	–	–	1	–	✓
50,51; 58,59/54,55; 62,63	1	1	–	–	1	–	✓
50,58; 54,62/51,55; 59,63	1	1	–	–	1	–	✓
52,53; 54,55/60,61; 62,63	1	1	–	1	–	–	✓
52,53; 60,61/54,55; 62,63	1	1	–	1	–	–	✓
52,54; 60,62/53,55; 61,63	1	1	–	1	–	–	✓
56,57; 58,59/60,61; 62,63	1	1	1	–	–	–	✓
56,57; 60,61/58,59; 62,63	1	1	1	–	–	–	✓
56,58; 60,62/57,59; 61,63	1	1	1	–	–	–	✓

Table B.7 (*continued*)

List (e)

	A	B	C	X	Y	Z	
8, 9; 10,11/40,41; 42,43//12,13; 14,15/44,45; 46,47	–	0	1	–	–	–	⎫
8, 9; 12,13/40,41; 44,45//10,11; 42,43/14,15; 46,47	–	0	1	–	–	–	⎬ S
8,10; 12,14/40,42; 44,46// 9,11; 13,15/41,43; 45,47	–	0	1	–	–	–	⎭
32,33; 34,35/36,37; 38,39//40,41; 42,43/44,45; 46,47	1	0	–	–	–	–	✓
32,33; 34,35/36,37; 38,39//48,49; 50,51/52,53' 54,55	1	–	0	–	–	–	✓
32,33; 34,35/40,41; 42,43//36,37; 38,39/44,45; 46,47	1	0	–	–	–	–	✓
32,33; 34,35/40,41; 42,43//48,49; 50,51/56,57; 58,59	1	–	–	0	–	–	✓
32,33; 34,35/48,49; 50,51//36,37; 38,39/52,53; 54,55	1	–	0	–	–	–	✓
32,33; 34,35/48,49; 50,51//40,41; 42,43/56,57; 58,59	1	–	–	0	–	–	✓
32,33; 36,37/42,43; 44,45//34,35; 38,39/44,45; 46,47	1	0	–	–	–	–	✓
32,33; 36,37/48,49; 52,53//34,35; 38,39/50,51; 54,55	1	–	0	–	–	–	✓
32,33; 36,37/48,49; 52,53//40,41; 56,57/44,45; 60,61	1	–	–	–	0	–	✓
32,33; 40,41/48,49; 56,57//34,35; 50,51/42,43; 58,59	1	–	–	0	–	–	✓
32,33; 40,41/48,49; 56,57//36,37; 52,53/44,45; 60,61	1	–	–	0	–	–	✓
32,34; 36,38/40,42; 44,46//33,35; 37,39/41,43; 45,47	1	0	–	–	–	–	✓
32,34; 36,38/40,42; 44,46//48,50; 52,54/56,58; 60,62	1	–	–	–	–	0	✓
32,34; 36,38/48,50; 52,54//33,35; 37,39/49,51; 53,55	1	–	0	–	–	–	✓
32,34; 36,38/48,50; 52,54//40,42; 56,58/44,60; 46,62	1	–	–	–	0	–	✓
32,34; 40,42/48,50; 56,58//33,35; 41,43/49,57; 51,59	1	–	–	0	–	–	✓
32,34; 40,42/48,50; 56,58//36,38; 52,54/44,60; 46,62	1	–	–	–	0	–	✓
32,36; 40,44/48,52; 56,60//33,37; 41,45/49,53; 57,61	1	–	–	–	0	–	✓
32,36; 40,44/48,52; 56,60//34,38; 50,54/42,46; 58,62	1	–	–	–	0	–	✓
32,40; 48,56/33,41; 49,57//34,35; 50,51/42,43; 58,59	1	–	–	0	–	–	✓
32,40; 48,56/33,41; 49,57//36,37; 52,53/44,45; 60,61	1	–	–	–	0	–	✓
33,35; 37,39/41,43; 45,47//49,51; 53,55/57,59; 61,63	1	–	–	–	–	1	⎫
33,35; 37,39/49,51; 53,55//41,43; 45,47/57,59; 61,63	1	–	–	–	–	1	⎬
33,35; 41,43/49,57; 51,59//37,39; 45,47/53,55; 61,63	1	–	–	–	–	1	⎬ ✓
33,37; 41,45/49,53; 57,61//35,39; 43,47/51,55; 59,63	1	–	–	–	–	1	⎭
34,35; 38,39/42,43; 46,47//50,51; 54,55/58,59; 62,63	1	–	–	–	1	–	⎫
34,35; 38,39/50,51; 54,55//42,43; 46,47/58,59; 62,63	1	–	–	–	1	–	⎬
34,35; 50,51/42,43; 58,59//38,39; 46,47/54,55; 62,63	1	–	–	–	1	–	⎬ ✓
34,38; 50,54/42,46; 58,62//35,39; 43,47/51,55; 59,63	1	–	–	–	1	–	⎭
36,37; 38,39/44,45; 46,47//52,53; 54,55/60,61; 62,63	1	–	–	1	–	–	⎫
36,37; 38,39/52,53; 54,55//44,45; 46,47/60,61; 62,63	1	–	–	1	–	–	⎬
36,37; 52,53/44,45; 60,61//38,39; 46,47/54,55; 62,63	1	–	–	1	–	–	⎬ ✓
36,38; 52,54/44,60; 46,62//37,39; 45,47/53,55; 61,63	1	–	–	1	–	–	⎭
40,41; 42,43/44,45; 46,47//56,57; 58,59/60,61; 62,63	1	–	1	–	–	–	⎫
40,41; 42,43/56,57; 58,59//44,45; 46,47/60,61; 62,63	1	–	1	–	–	–	⎬
40,41; 56,57/44,45; 60,61//42,43; 46,47/58,59; 62,63	1	–	1	–	–	–	⎬ ✓
40,42; 56,58/44,60; 46,62//41,43; 45,47/57,59; 61,63	1	–	1	–	–	–	⎭
48,49; 50,51/52,53; 54,55//56,57; 58,59/60,61; 62,63	1	1	–	–	–	–	⎫
48,49; 50,51/56,57; 58,59//52,53; 54,55/60,61; 62,63	1	1	–	–	–	–	⎬
48,49; 52,53/56,57; 60,61//50,51; 54,55/58,59; 62,63	1	1	–	–	–	–	⎬ ✓
48,50; 52,54/56,58; 60,62//49,51; 53,55/57,59; 61,63	1	1	–	–	–	–	⎭

List (f)

	A	B	C	X	Y	Z	
32,33; 34,35/36,37; 38,39//40,41; 42,43/44,45; 46,47 48,49; 50,51/52,53; 54,55//56,57; 58,59/60,61; 62,63	1	–	–	–	–	–	⎫ T
32,33; 34,35/36,37; 38,39//48,49; 50,51/52,53; 54,55 40,41; 42,43/44,45; 46,47//56,57; 58,59/60,61; 62,63	1	–	–	–	–	–	⎬
32,33; 34,35/40,41; 42,43//48,49; 50,51/56,57; 58,59 36,37; 38,39/44,45; 46,47//52,53; 54,55/60,61; 62,63	1	–	–	–	–	–	⎬
32,33; 36,37/48,49; 52,53//40,41; 56,57/44,45; 60,61 34,35; 38,39/42,43; 46,47//50,51; 54,55/58,59; 62,63	1	–	–	–	–	–	⎬
32,34; 36,38/40,42; 44,46//48,50; 52,54/56,58; 60,62 33,35; 37,39/41,43; 45,47//49,51; 53,55/57,59; 61,63	1	–	–	–	–	–	⎭

B.5 Tabular minimisation of six variable functions

An example of tabular minimisation of a six variable function will be given in Example B.3.

Example B.3
Minimise the function represented in the numerical sum-of-products form by

$$F = 4, 6, 7, 8, 9, 10, 11, 12, 13, 14, 15, 32, 33, 34, 35, 36, 37, 38, 39, 40, 41, 42, 43,$$
$$44, 45, 46, 47, 48, 49, 50, 51, 52, 53, 54, 55, 56, 57, 58, 59, 60, 61, 62$$
and 63.

Solution
To simplify the function using the tabular method, the function is first expressed in binary form (Table B.6), and arranged into groups according to the number of uncomplemented digits contained in each term, as shown in Table B.7, list (a). The function is then tabulated, and the prime implicants are determined as shown in Table B.7, lists (b) to (f). The results obtained are summarised below.

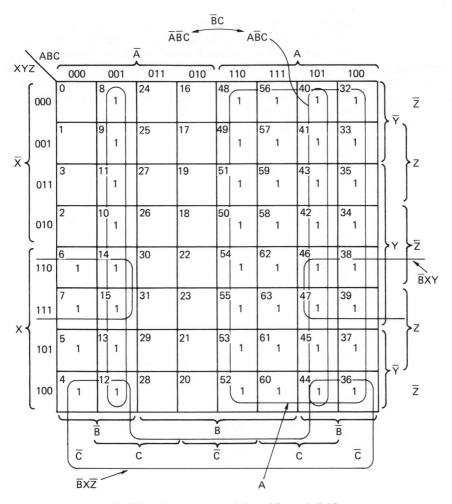

Fig B 7   Karnaugh map solution of Example 5.13

From list (d),

> P is given by 4, 6, 12, 14, 36, 38, 44 and 46, or $P = \bar{B}X\bar{Z}$
> Q is given by 8, 9, 10, 11, 12, 13, 14 and 15, or $Q = \bar{A}\bar{B}C$
> R is given by 6, 7, 14, 15, 38, 39, 46 and 47, or $R = \bar{B}XY$

From list (e),

> S is given by 8, 9, 10, 11, 12, 13, 14, 15, 40, 41, 42, 43, 44, 45, 46 and 47, or
> $S = \bar{B}C$

Finally, from list (f),

> T is given by 32, 33, 34, 35, 36, 37, 38, 39, 40, 41, 42, 43, 44, 45, 46, 47, 48, 49,
> 50, 51, 52, 53, 54, 55, 56, 57, 58, 59, 60, 61, 62 and 63, or $T = A$

To summarise, the prime implicants are $P = \bar{B}X\bar{Z}$, $Q = \bar{A}\bar{B}C$, $R = \bar{B}XY$, $S = \bar{B}C$ and $T = A$. Since Q is covered by S, the simplified version of the function F is given by $F = P + R + S + T$ or, in terms of the variables $F = \bar{B}X\bar{Z} + \bar{B}XY + \bar{B}C + A$. Note that there was no need to draw a prime implicants matrix; the final simplification was made by inspection of the prime implicants listed above.

The function F, given originally in numerical form, is now plotted on a six-variable Karnaugh map, Fig. B.7, with reference to Table B.6, for comparison with the above. Since all the lists of the tabular method, and the Boolean algebra expressions used, employ the alphabetic order of the variables, i.e. the variable A is treated as the most significant digit, the Karnaugh map is now also marked with A as the most significant digit.

From the Karnaugh map of Fig. B.7, the simplified function F is given by $F = A + \bar{A}\bar{B}C + \bar{B}XY + \bar{B}X\bar{Z}$. Since A also covers $A\bar{B}C$, the function F reduces to $F = A + \bar{B}C + \bar{B}XY + \bar{B}X\bar{Z}$, which is the same as the one obtained by the tabular method.

Appendix C

DIGITAL CLOCK DESIGN

The circuits described in this appendix can be constructed fairly easily with a minimum amount of effort and wiring using a breadboard such as one of those described in Chapter 11.

A number of complete circuits of digital clocks with various refinements are available in integrated circuit form, on a single chip, but there is no substitute for an experience of designing and building a system using the 7490 decade counters, or similar ICs and other associated components.

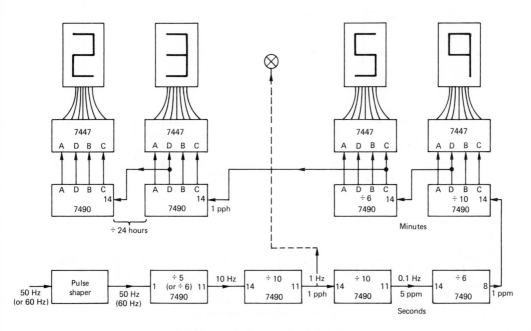

Fig C 1   Block diagram of a digital clock

It is assumed that the synchronising signal for the clock is derived from the 50 Hz or 60 Hz mains supply, together with the 5 V, V_{CC} supply required for the complete clock circuit. A signal generator and power supply unit can be used in the laboratory to provide these functions if preferred.

To simplify the description of the design, only a block diagram of the complete clock is given in Fig. C.1. with the necessary connections for either 24 or 12 hour operation in Figs. C.2 and C.3 respectively.

A signal of 4 V to 5 V is derived from the mains transformer, providing the supply to the rest of the system, and is applied to a pulse shaper. This could be a two transistor, or an IC Schmitt trigger circuit. The output of this circuit, which is a square wave, is divided by 5 or 6, depending on the mains frequency, to obtain 10 Hz. This is further divided by 10 to give 1 Hz, or one pulse per second. Further division by 60 gives one pulse per minute.

Although no provision is shown in Fig. C.1 for displaying seconds, this could easily be achieved by simply connecting the A, B, C and D outputs of the divide by 10 and divide by 6 units via seven segment decoders and drivers to feed two additional seven segment displays. The connections are similar to those shown for the minutes and hours.

The two possible versions for displaying the time, i.e. the 24 hour and 12 hour, are shown

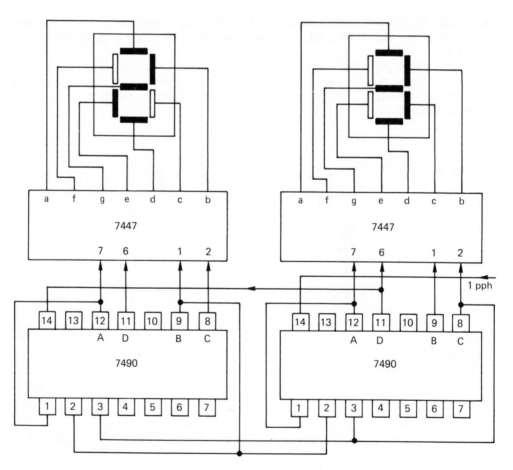

Fig C 2   24 hour counter

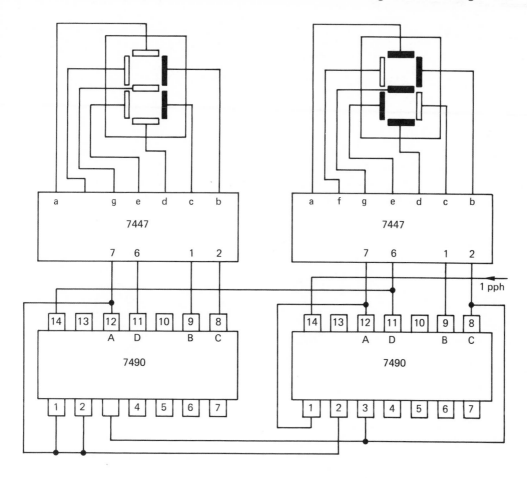

Fig C 3 12 hour counter

with the necessary connections in Figs. C.2 and C.3 respectively. Instead of the seconds display a simplified arrangement is incorporated in the form of a light emitting diode switching on and off at a rate of one pulse per second. A current limiting resistor should be connected in series with the LED or, perhaps a better solution, an inverter in the form of a buffer amplifier should be used as well.

A quick time setting arrangement is provided. This is done by means of gating hours and minutes at a rate of one pulse per second and 10 pulses per second respectively for fast setting, and feeding the input to minutes at a rate of one pulse per second for the slow setting. To avoid switch bounce problems, cross coupled NAND gates should be used in the arrangement as described in Chapter 7.

Having built and experimented with the above described digital clock it should not be too difficult to adapt any part of the circuit to suit possible new requirements.

INDEX